AIR CONDITIONING: Procedures and Installation

V. PAUL LANG

CBS

CBS PUBLISHERS & DISTRIBUTORS
NEW DELHI • BANGALORE

Preface

AIR CONDITIONING: PROCEDURES AND INSTALLATION is a procedures manual which provides basic instruction in installing air-conditioning units and gas-fired furnaces. Step-by-step procedures are given for residential equipment installation, maintenance, and troubleshooting.

The first units cover the areas of required local and national codes and ordinances in terms of safety (proper operation of equipment); tools for installing and maintaining equipment including refrigerant tubing procedures; explanations and applications of basic electrical terms, symbols, wire characteristics, and sizing; control checkout and troubleshooting; and preparations for installation. Later units give the actual procedures for installing central residential cooling, heating, and heating and cooling (heat pump) systems. Electronic air cleaners and humidifiers help complete the total system.

This text expands upon the basic theory covered in a companion text, PRINCIPLES OF AIR CONDITIONING. Procedures can apply to any type of air-conditioning product. The information presented is intended to prepare the student or apprentice for entry-level positions in the field of air-conditioning installation and maintenance.

V. Paul Lang has been associated with the air-conditioning industry for many years. Formerly he was a member of the Carrier Air Conditioning Company Application Engineering Department, and prior to that was employed by the Minneapolis Honeywell Company. Presently, Mr. Lang is manager of the technical writing services at Carrier. Mr. Lang received his undergraduate degree at St. John's University, Collegeville, Minnesota and completed graduate master's credits at the University of Minnesota.

All Delmar materials are as error free as possible. Please offer any suggestions or corrections by writing to the Editor in Chief at the address below.

Delmar also publishes the following titles for Air Conditioning/Refrigeration.

PRINCIPLES OF AIR CONDITIONING
REFRIGERATION SERVICING
PRINCIPLES OF REFRIGERATION
PRACTICAL PROBLEMS IN MATHEMATICS FOR HEATING AND COOLING TECHNICIANS

ISBN : 81-239-1167-X

First Indian Edition : 1987
Reprint : 1998
Reprint : 2004

Published by : Satish Kumar Jain for CBS Publishers & Distributors,
4596/1-A, 11 Darya Ganj, New Delhi - 110 002 (India)

Printed at : Asia Printograph, Shahdara, Delhi - 110 032 (India)

Contents

Unit 1 **Codes and Definitions** 1

Codes, ordinances, and standards – Definitions

Unit 2 **Standard Tools for Installing and Maintaining Air-conditioning Equipment** 9

Introduction – Hand tools – Refrigerant piping tools – Electric meters – Airside tools

Unit 3 **Refrigerant Tubing, Tools, and Procedures** 27

Introduction – Working with copper tubing – Tube bending – Tube cutting – Common refrigerant piping fittings – Tubing summary – Soldering

Unit 4 **Electrical Terms and Symbols, and Wire Characteristics and Sizing** 43

Basic electrical terms – Electrical symbols – Characteristics of electric wire – How to select wire size

Unit 5 **Control Checkout and Troubleshooting** 53

Introduction – General checkout procedures – Troubleshooting the system

Unit 6 **Preparing to Install the Air-conditioning Unit** 63

Before installation begins – Base pads for outdoor units

Unit 7 **Installing Room Air Conditioners** 69

Introduction – Step 1 - Getting ready to install the unit – Step 2 - Test the power source – Step 3 - Install the unit – Step 4 - Start up the unit – Step 5 - Measure the discharge air temperature – Step 6 - Measure return air temperature – Step 7 - Clean up the job site – Step 8 - Give the instructions to the owner – General servicing rules and techniques – General service problems

Unit 8 **Installing Central Residential Cooling Coils** 77

Introduction – Step 1 - Getting ready to install the coil – Step 2 - Install the coil and casing – Step 3 - Start up the air-conditioning system – Step 4 - Measure the discharge air temperature – Step 5 - Measure the return air temperature – Step 6 - Clean up the job site – Step 7 - Give the instructions to the owner – General service problems

Unit 9 **Installing Central Residential Cooling Systems** 85

Introduction – Step 1 - Getting ready to install the outdoor condensing unit – Step 2 - Mount the outdoor base pad and condensing unit – Step 3 - Mount the cooling coil in the furnace plenum – Step 4 - Install the refrigerant piping – Step 5 - Install the wiring – Step 6 - Check out all piping and wiring – Step 7 - Start up the system – Step 8 - Clean up the job site – Step 9 - Instruct the owner in operation and maintenance – General service problems

Unit 10 **Installing Two-piece Residential Heat-pump Systems** 95

Introduction – Step 1 - Getting ready to install the outdoor unit – Step 2 - Mount the outdoor base pad and condensing unit – Step 3 - Mount the indoor unit coil and fan – Step 4 - Install the refrigerant piping – Step 5 - Install the wiring – Step 6 - Check out all piping and wiring – Step 7 - Start up the system – Step 8 - Clean up the job site – Step 9 - Instruct the owner in operation and maintenance – General service problems

Unit 11 **Installing Gas-fired Residential Furnaces** 105

Introduction – Step 1 - Getting ready to install the furnace – Step 2 - Make duct connections to the furnace – Step 3 - Install the furnace flue – Step 4 - Install the gas piping – Step 5 - Install the wiring – Step 6 - Check out all piping and wiring – Step 7 - Start up the furnace – Step 8 - Clean up the job site – Step 9 - Instruct the owner in operation and maintenance – General service problems – Combustion air

Unit 12 **Installing Electronic Air Cleaners** 113

Introduction – Step 1 - Getting ready to install the electronic air cleaner – Step 2 - Prepare the furnace and install the air cleaner – Step 3 - Install wiring to the electronic air cleaner – Step 4 - Clean up the job site – Step 5 - Instruct the owner in operation and maintenance

Unit 13 **Installing Central Residential Humidifiers** 117

Introduction – Step 1 - Getting ready to install the humidifier – Step 2 - Prepare the supply air plenum duct and install the humidifier – Step 3 - Install the water line to the humidifier – Step 4 - Install the wiring and check humidifier operation – Step 5 - Clean up the job site – Step 6 - Instruct the owner in operation and maintenance

Answers to Review 122

Index 123

1
Codes and Definitions

OBJECTIVES

After studying this unit, the student will be able to

- list sources of industry-related codes, ordinances, and standards.
- describe the basic components and accessories of residential heating and cooling products.

CODES, ORDINANCES, AND STANDARDS

National and local codes, local ordinances, and national industry association standards are well established. They are useful guides for the proper installation and performance of refrigeration and air-conditioning products and systems. Each code, ordinance, and standard has a defined function. As a whole, they promote high-level product performance, system efficiency, and personal safety.

Codes and Ordinances

Codes and ordinances are usually national or local in origin. (Local means state, town, county, city, and village.) They include national and local safety codes and local heating, plumbing, and building codes. Sound-level or noise ordinances are also included. Since air-conditioning installations must conform to these codes and ordinances, the technician should be familiar with the regulations that apply in his or her locality.

Standards

Standards that range from the application of basic principles to product performance are set by various industry-oriented societies and institutes. Some long-established, reliable sources and their general areas of activity are:

Underwriters' Laboratories, Inc. (UL), 333 Pfingsten Road, Northbrook, Illinois 60062. Appliances are UL approved only after passing electric-safety tests and code requirements set by Underwriters' Laboratories.

American Gas Association (AGA), 1515 Wilson Boulevard, Arlington, Virginia 22209. Gas-fired products (such as furnaces) must meet AGA safety and performance standards before they can display the AGA seal.

American Refrigeration Institute (ARI), 1815 North Fort Meyer Drive, Arlington, Virginia 22209. An industry-sponsored group, the Institute establishes standards for refrigeration and air-conditioning products. The ARI seal is issued only after certification against these standards.

American Society of Heating, Refrigeration, and Air-Conditioning Engineers (ASHRAE), 345 East 47th Street, New York, New York 10017. ASHRAE affects or influences most aspects of the industry, but its most popular contribution is the ASHRAE Guide and related reference handbooks.

The American National Standards Institute (ANSI) in New York, New York; the National Environmental Systems Contractors (NESC) in Arlington, Virginia; and the National Fire Protection Association (NFPA) in Boston, Massachusetts are among several additional organizations that directly influence standards in the industry. The National Fire Protection Association publishes the National Electrical Code. In addition, most manufacturers of air-conditioning products furnish literature that describes the proper application of their products.

DEFINITIONS

Before learning to install and service air-conditioning units, the student must be familiar with the following basic components and accessories used in residential heating and cooling systems.

Cooling

Packaged unit. This packaged unit provides cooling only. It contains a cooling coil, condensing coil, and compressor in a one-piece unit. It may be installed through the wall, roof, or outdoors as shown in figure 1-1. All components are factory-connected and usually require only electrical and duct hookups, and installation of drain lines.

Cooling coil. Also called the *evaporator coil*, it is one of the three basic components in a central residential system. Usually the cooling coil is installed inside the residence in the furnace plenum, and is connected by refrigerant tubing to the condensing unit outside the residence. An alternate method is to install the cooling coil in the attic duct run. The furnace fan section blower moves room air across the coil tubes where it is cooled, figure 1 2.

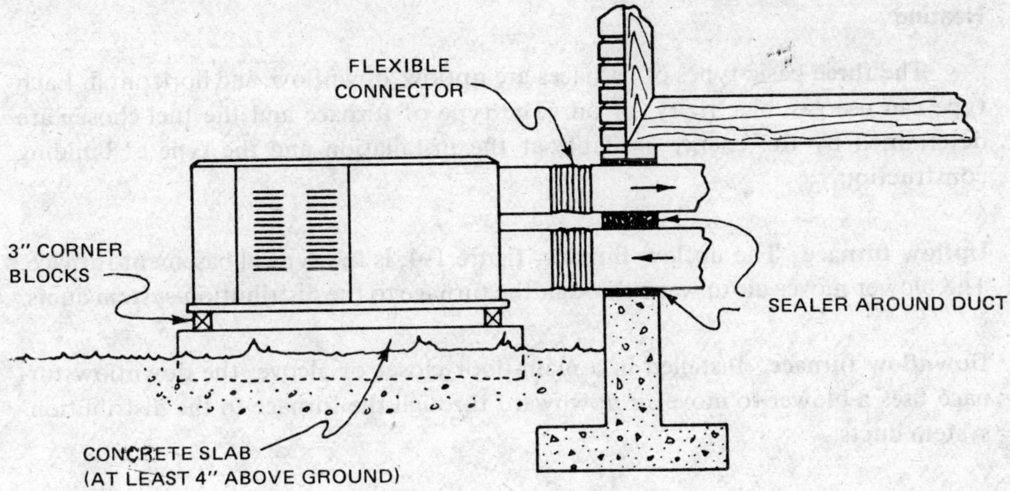

Fig. 1-1 Typical one-piece cooling unit installation

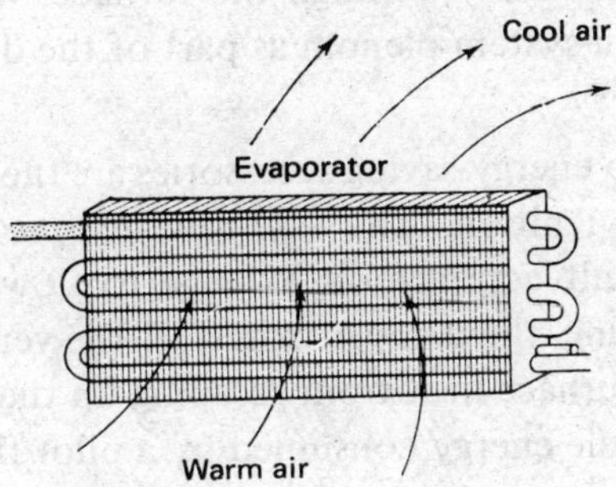

Fig. 1-2 Typical cooling coil

Condensing unit. The compressor and the condensing coil are combined in a single shell or cabinet to form the condensing unit, figure 1-3. The *compressor* pumps, compresses, and heats the refrigerant from the cooling coil. The unit then passes the refrigerant to the condensing coil. The *condensing coil fan* cools the hot refrigerant in the coil tubes by blowing or drawing outside air across the tubes. As a result, refrigerant heat is transferred to the outside air.

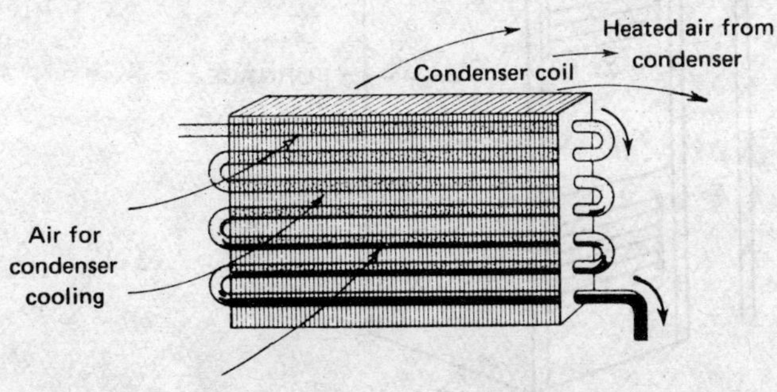

Fig. 1-3 Typical condenser coil

Heating

The three basic types of furnaces are upflow, downflow, and horizontal. Each type can use gas, electricity, or oil. The type of furnace and the fuel chosen are determined by the energy available at the installation and the type of building construction.

Upflow furnace. The upflow furnace, figure 1-4, is the typical basement furnace. The blower moves air upward through the furnace to the distribution-system ducts.

Downflow furnace. Installed in a main-floor closet or alcove, the downflow furnace uses a blower to move air downward through the furnace to the distribution-system ducts.

Horizontal furnace. The horizontal furnace is located in the attic or crawl space. A blower moves air horizontally through the furnace, which is usually installed directly in the distribution-system plenum as part of the ductwork.

Furnace accessories. Two energy-saving accessories are the furnace vent-pipe damper and automatic electric ignition. The *vent-pipe damper* operates in conjunction with the furnace to permit combustion exhaust to flow to the chimney. When the furnace is not operating, the damper closes off the vent pipe. This action prevents built-up or stored furnace heat from escaping up the chimney. The *automatic electric ignition* saves the energy consumed by a pilot flame, because it operates only when the main burner must be lit.

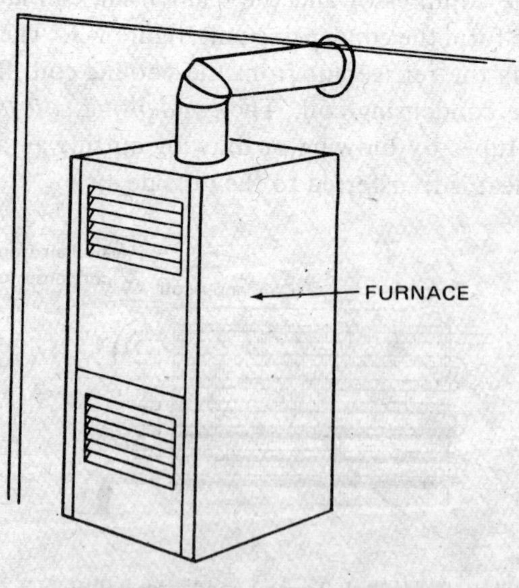

Fig. 1-4 Typical upflow gas furnace

Heat Pumps

Packaged unit. Designed to provide both cooling and heating, a packaged heat pump contains the cooling coil, compressor, and condensing coil in one shell or cabinet. A four-way valve makes both heating and cooling possible. Set in position ONE, the valve directs refrigerant to the compressor in the typical cooling-circuit flow pattern. When set in position TWO, the valve directs refrigerant to the compressor in the opposite, heating-circuit, flow pattern. In other words, the cooling coil and condensing coil exchange functions as dictated by the valve setting.

Two-piece heat pump. In the two-piece heat pump, figure 1-5, the condensing unit (condensing fan coil, compressor, and four-way valve) is located outdoors. The indoor section contains the cooling coil. If a furnace is not part of the system, a fan-coil unit (a cooling coil with a fan) is used indoors.

Three-piece heat pump. This product is the same as the two-piece unit, except the compressor and four-way valve are separated from the outdoor section and located indoors in a third cabinet, usually situated in the garage.

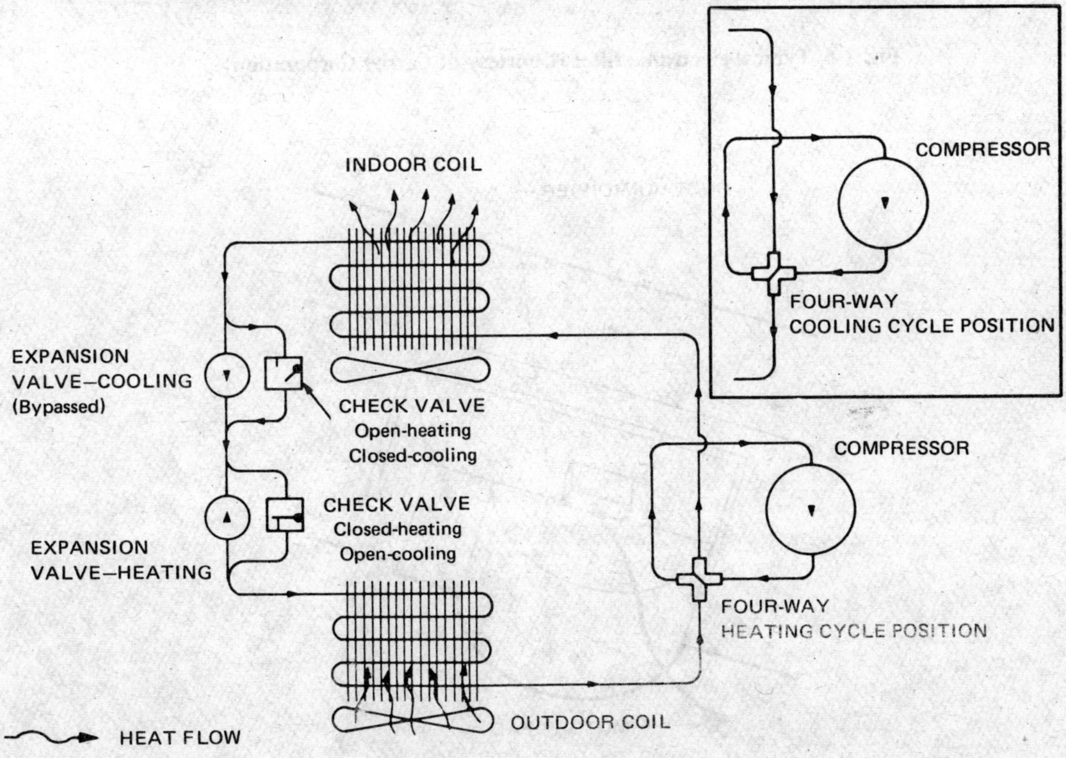

Fig. 1-5 Typical two-piece heat pump

Fig. 1-6 Typical electronic filter (Courtesy of Carrier Corporation)

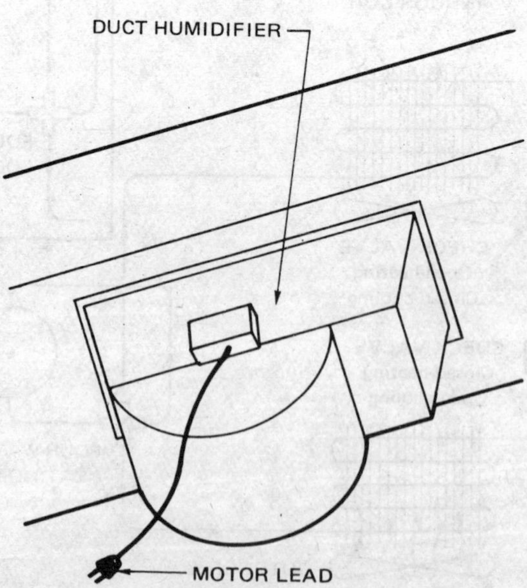

Fig. 1-7 Typical duct humidifier

Filters

Electronic or mechanical filters are installed in the air-distribution duct system, usually in front of the furnace or cooling unit. Electronic filters, figure 1-6, remove smaller dirt particles. Thus, they create cleaner air than a typical mechanical filter.

Humidifiers

Free-standing, furnace-plenum, or duct-installed humidifiers, figure 1-7, add moisture to a residence during periods of dryness. Duct-installed and furnace-plenum units are part of a central system and add moisture to the whole house. Free-standing units are usually effective in only a limited area in the house.

REVIEW

Match the items in Column I with the best choice from Column II.

Column I

1. Codes, ordinances, and standards
2. Cooling coil
3. Heat pump
4. Condensing unit
5. Furnace
6. Underwriters' Laboratories
7. Vent-pipe damper
8. Filters
9. Automatic ignition
10. American Society of Heating, Refrigeration, and Air-Conditioning Engineers

Column II

a. ASHRAE Guide
b. Evaporator coil
c. Can be upflow, downflow, or horizontal
d. Electric safety codes
e. Prevents heat from escaping
f. Electronic or mechanical
g. Condensing coil and compressor in a single cabinet
h. Eliminates the need for a pilot flame on a furnace
i. Performs both cooling and heating functions
j. Influence performance, efficiency, and safety

2
Standard Tools for Installing and Maintaining Air-conditioning Equipment

OBJECTIVES

After studying this unit, the student will be able to

- demonstrate the use and care of tools needed to install and maintain air-conditioning and refrigeration equipment.

INTRODUCTION

Hand, refrigerant-piping, electric, pressure-measuring, and air-measuring and balancing tools are the major implements used in air-conditioning installation and service. In some cases the name of the tool describes its operation. For example, a packing wrench is used to tighten packing nuts.

High-quality tools, though more costly, are a sound investment. A tool that does not become distorted and which remains accurate after repeated use is a "good buy." Some typical tools and their application to the air-conditioning trade include:

- Ratchet and key wrenches for service valves
- Setting hammer to form sheetmetal
- Allen setscrew wrenches for replacing or adjusting motor pulleys
- Pinch-off tool to pinch off tubing
- Pocket thermometer for measuring temperature
- Ohmmeter to determine electrical continuity
- Halide leak detector to find leaks
- Airflow meter to measure air speed

Many of these and other related tools are discussed in this unit.

HAND TOOLS

Hammers, wrenches, pliers, screwdrivers, chisels, and hand drills are direct extensions of the hand and are therefore termed *hand tools*. As part of their maintenance routine, all metal surfaces of these tools should periodically be lightly oiled.

Hammers

Nail hammer. The nail hammer, figure 2-1, is used to drive or pull nails. Before using it, always inspect the hammer to be sure the head is securely fastened to the handle. If the head is loose, drive tightening wedges in the head end of the handle to secure the head.

When driving nails, always strike with the face of the hammer; it is hardened for that purpose. To pull nails, insert the claws of the hammer under the head of the nail and pull until the handle is nearly vertical. Then place a wedge under the hammer head and pull until the nail is removed, figure 2-2.

Setting hammer. Use a setting hammer for striking and forming sheet metal, as shown in figure 2-3. Wear safety goggles when using this tool.

Ball peen hammer. The ball peen hammer, figure 2-4, is designed for heavy use and for forming heavy metals. Always inspect the hammer to be sure the head is firmly secured. Always wear safety goggles when using this tool.

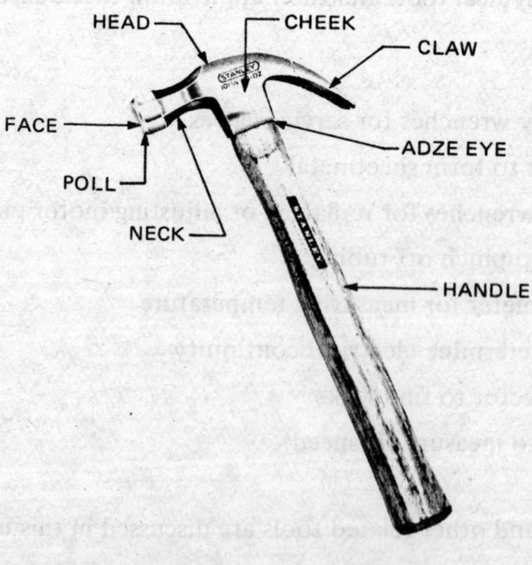

Fig. 2-1 Nail hammer

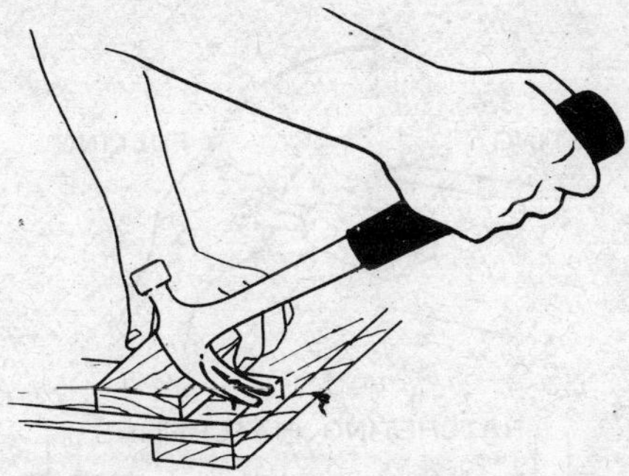

Fig. 2-2 Pulling nails

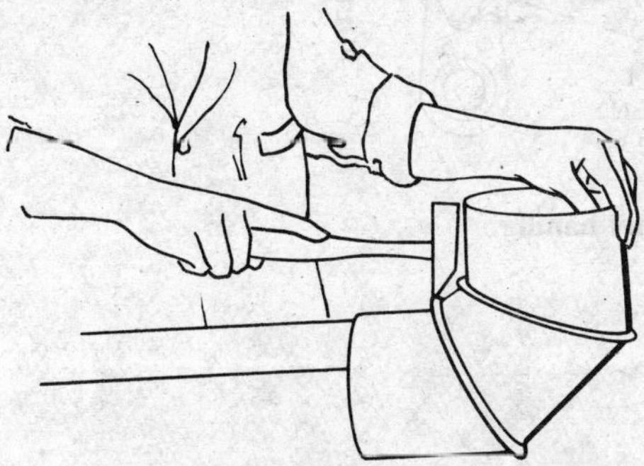

Fig. 2-3 Setting hammer

Fig. 2-4 Ball peen hammer (Courtesy of Stanley Tools)

Wrenches

The wide variety of wrenches available enables the worker to choose the right tool for the specific task. When using any wrench, the safe practice is to pull on the handle rather than to push against it.

Socket wrenches. These wrenches are used when there is free access to the nut to be removed. Socket wrenches are available in varying sizes, figure 2-5, and with

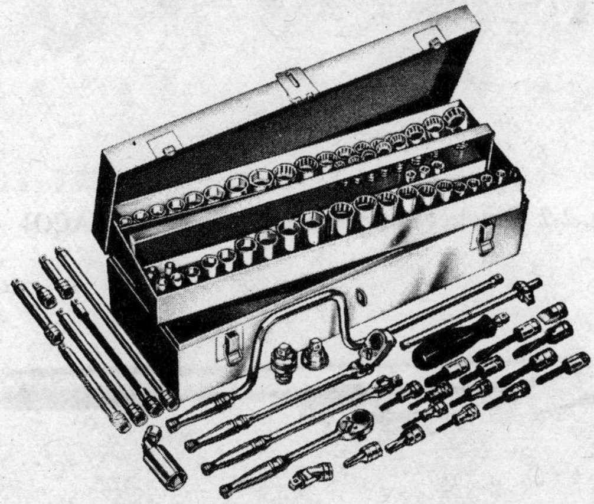

Fig. 2-5 Socket wrench set (Courtesy of Snap-on Tools)

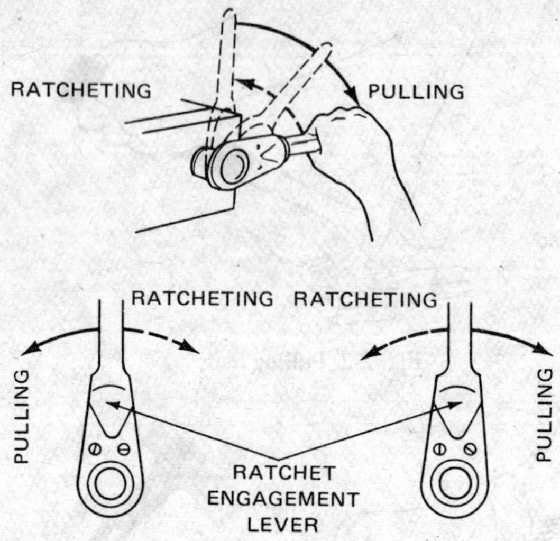

Fig. 2-6 Ratchet handle

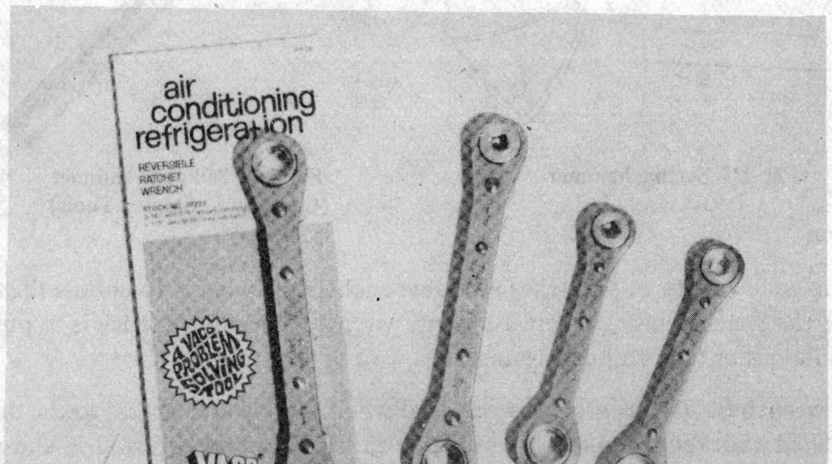

Fig. 2-7 Ratchet box wrench (Courtesy of VACO)

Fig. 2-8 Torque wrench

swivel, ratchet, torque, or speed handles. Ratchet handles, figure 2-6, allow complete or partial tightening or loosening of the nut.

Ratchet box wrench. A ratchet box wrench, figure 2-7, is used when there is only limited space around the nut in which to move the wrench handle. This type of wrench is available in several sizes.

Torque wrench. Torque wrenches, figure 2-8, are used in cases where nuts, bolts, gaskets, and assembled parts require measured tightness. A measurement scale or dial, figure 2-9, or a signal light is usually built into the handle, providing a convenient way of reading the pressure being applied to the part.

Box wrench. The open-end box wrench, figure 2-10, is used when the area around the nut is limited. It is also used when a socket or close-jawed wrench cannot be placed over the nut. Box wrenches are available in several sizes and styles.

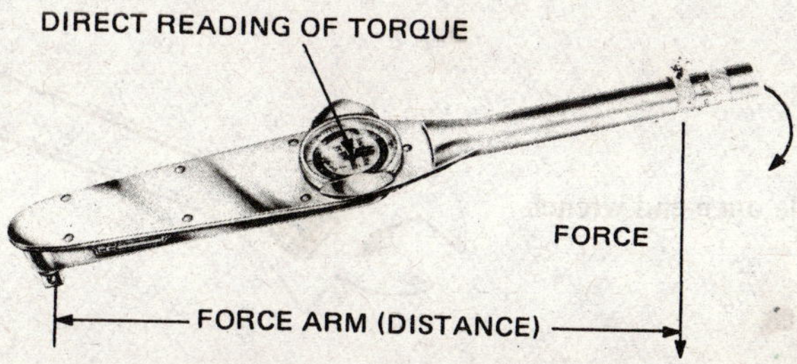

Fig. 2-9 Direct-reading torque wrench

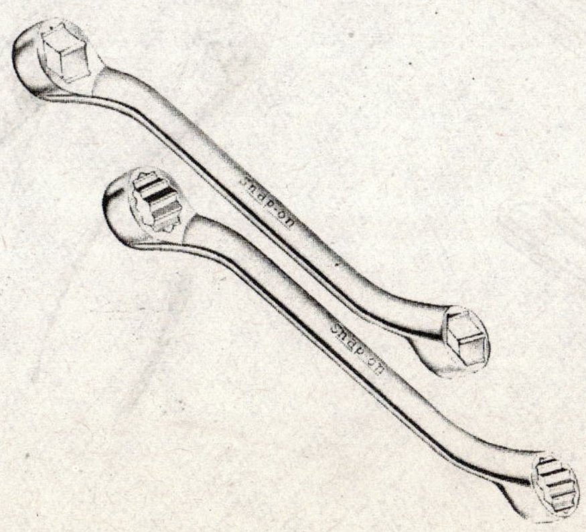

Fig. 2-10 Box wrenches

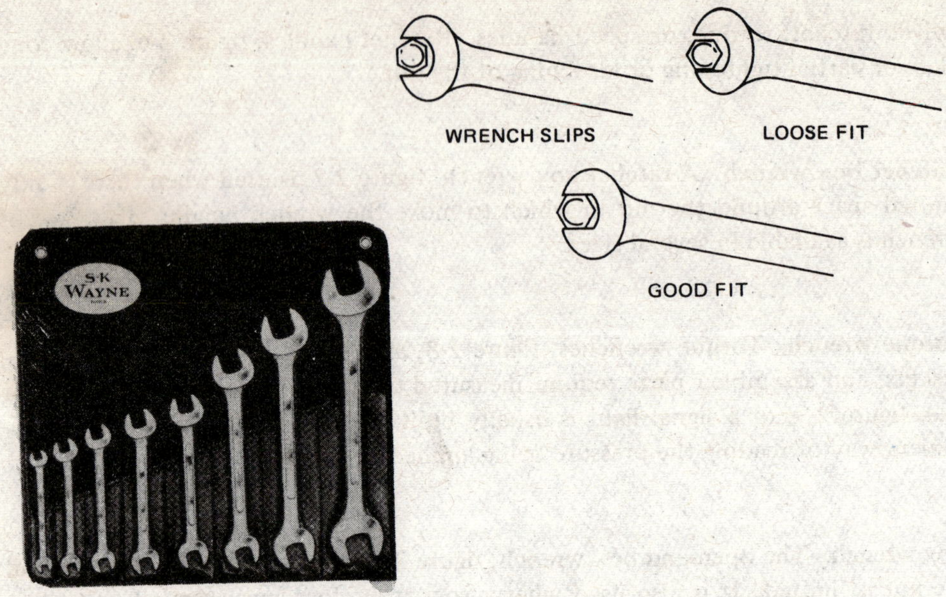

Fig. 2-11 Open-end wrenches

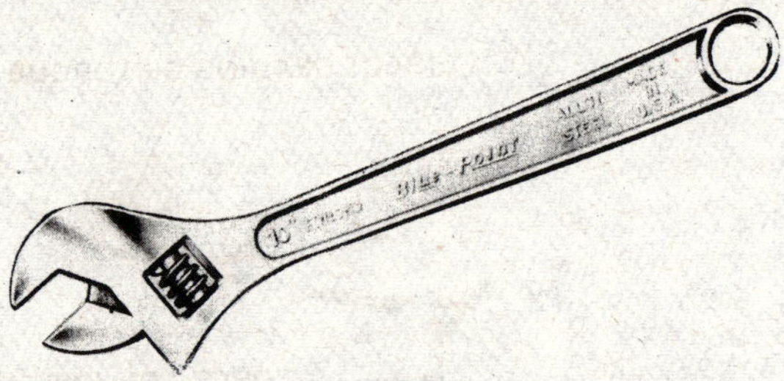

Fig. 2-12 Adjustable open-end wrench

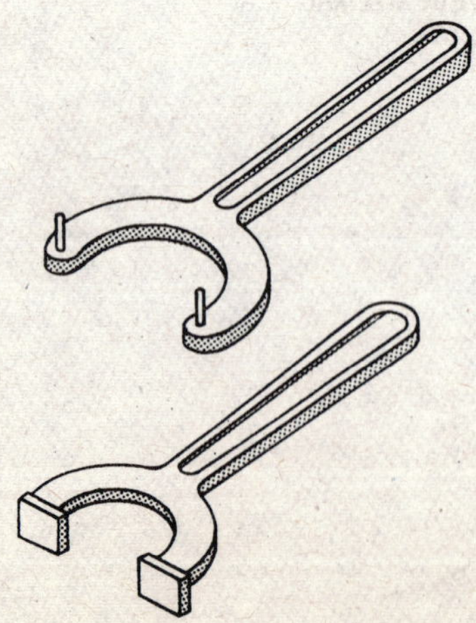

Fig. 2-13 Spanner wrenches

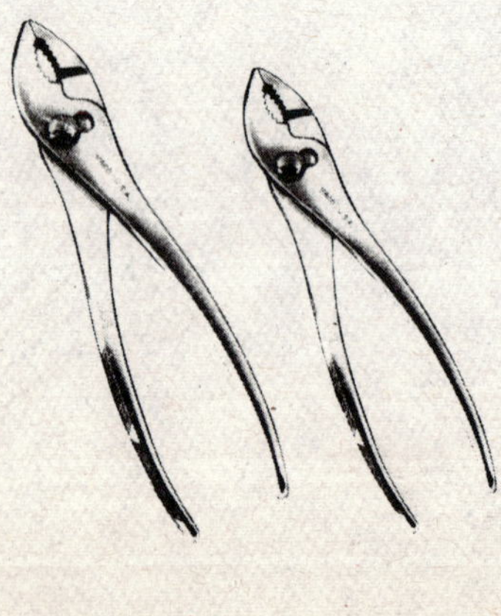

Fig. 2-14 Combination pliers

Open-end wrench. Any open-end wrench, figure 2-11, may be used where space is limited and it is not possible to use a socket wrench or a closed-end wrench. For example, open-end wrenches are useful when the nut to be removed or tightened is part of a continuous run of pipe or tubing.

Adjustable open-end wrench. This wrench, figure 2-12, is used in the same manner as any other open-end wrench. It has adjustable jaws that make it convenient for use on standard and nonstandard nuts. This type of wrench should be inspected before each use. When it is used properly, it can replace several sizes of nonadjustable wrenches.

Spanner wrench. Slotted or notched nuts require specially designed spanner wrenches, figure 2-13. One type of spanner wrench resembles a horseshoe with grips that fit into the slotted nut. A second type also resembles a horseshoe, but has pin grippers to fit the notched nut.

Pliers

Combination pliers. This type of pliers, figure 2-14, has adjustable jaws and is used to grip, cut, and hold small- and large-diameter objects.

Insulated side-cutter pliers. These pliers, figure 2-15, are used when cutting and gripping electrical wires. The handles are insulated to prevent electric shock when the pliers are properly used.

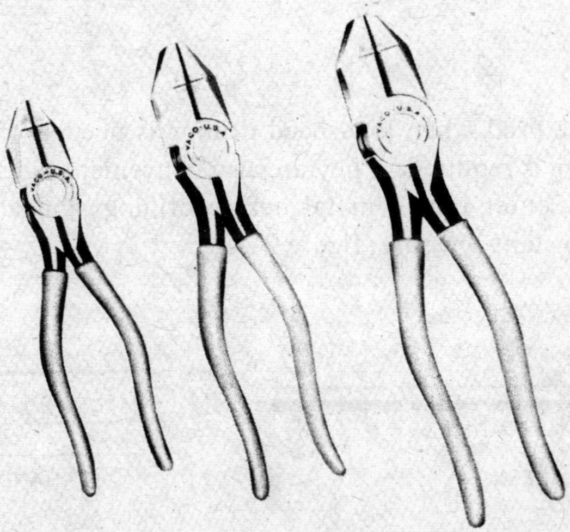

Fig. 2-15 Side-cutter pliers

Screwdrivers

Standard blade screwdriver. The commonly used standard blade screwdriver, figure 2-16, has a wooden or plastic handle that may or may not be insulated. The blade is made of a good grade of steel. It is securely anchored in the handle at the ferrule. This type of screwdriver is used to drive screws.

When using a screwdriver, select one fitted to the job to be done. Screwdrivers are specified by the length of the blade. The tip should be straight and the sides nearly parallel. In addition, the tip should fit the screw slot; it should not be wider than the screw head, figure 2-17.

Phillips blade screwdriver. The Phillips screwdriver, figure 2-18, is used to drive Phillips head screws. The tip of the Phillips blade is designed to match the crossed slots in the Phillips screw head. In all other respects, this driver is the same as the standard blade screwdriver.

Right-angle screwdriver. A right-angle screwdriver is required when limited space above the screw head prevents the use of a straight blade driver. The right-angle driver requires only a minimal space above the screw head.

Chisels

A standard wood chisel is convenient for making special cuts. Always keep your hands behind the cutting edge, figure 2-19, and, for best performance, keep the chisel properly sharpened.

Hand Drills

Hand drills can be used when low-speed drilling is needed or when special, difficult-to-reach drilling is required. They are also convenient where electricity is not readily available. Center punch metal before drilling, and always turn the crank with a steady, constant pressure, figure 2-20.

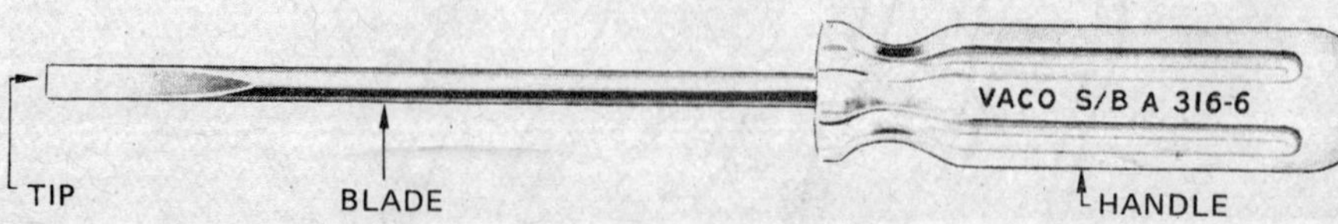

Fig. 2-16 Standard blade screwdriver

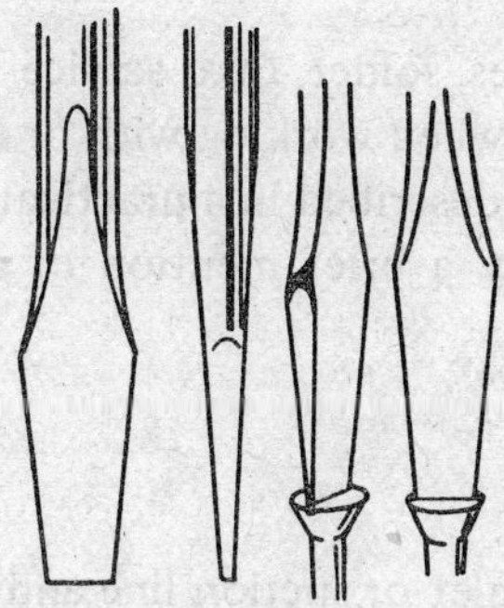

Fig. 2-17 Use the proper screwdriver.

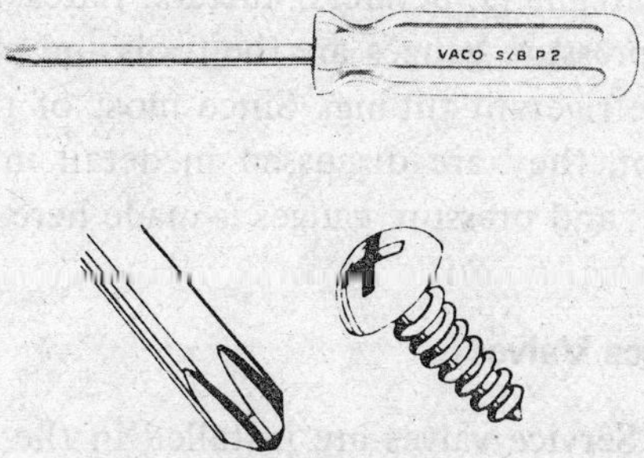

Fig. 2-18 Phillips head screwdriver and screw

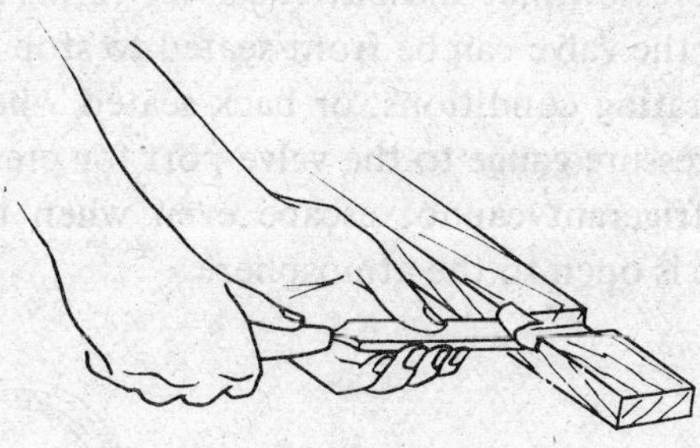

Fig. 2-19 Keep hands behind the cutting edge of the chisel.

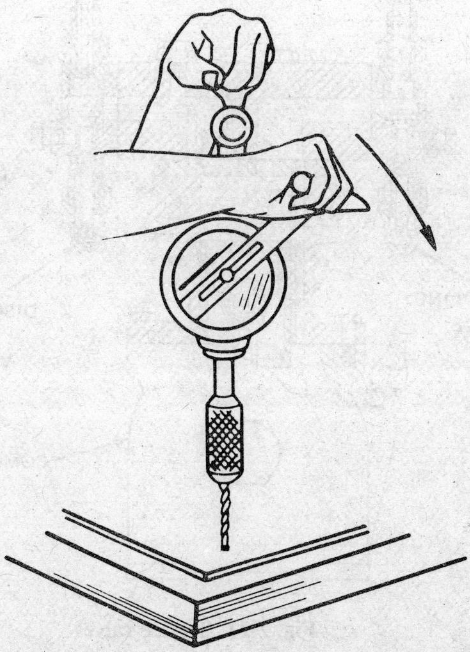

Fig. 2-20 Using a hand drill to drill metal

REFRIGERANT PIPING TOOLS

Reamers, benders, cutters, flares, swages, torches, solder, flux, service valves, and pressure gauges are the tools and materials used when working with or installing refrigerant tubing. Since most of these are best described in a practical application, they are discussed in detail in Unit 5. Only a brief mention of service valves and pressure gauges is made here.

Service Valves

Service valves are installed in the compressor inlet or suction line and in the compressor outlet or discharge line, as shown in figure 2-21. They are used when measuring pressure conditions in the system and when adding or charging refrigerant into the system. In addition, they may be used to remove or evacuate refrigerant.

Both the compressor inlet and outlet service valves are front-seated and back-seated. This means the valve can be front-seated to stop refrigerant flow, open for certain normal operating conditions, or back-seated when removing the pipe plug and connecting a pressure gauge to the valve port for pressure testing. In the back-seated position, refrigerant cannot escape even when the valve port plug is removed and the port is open to the atmosphere.

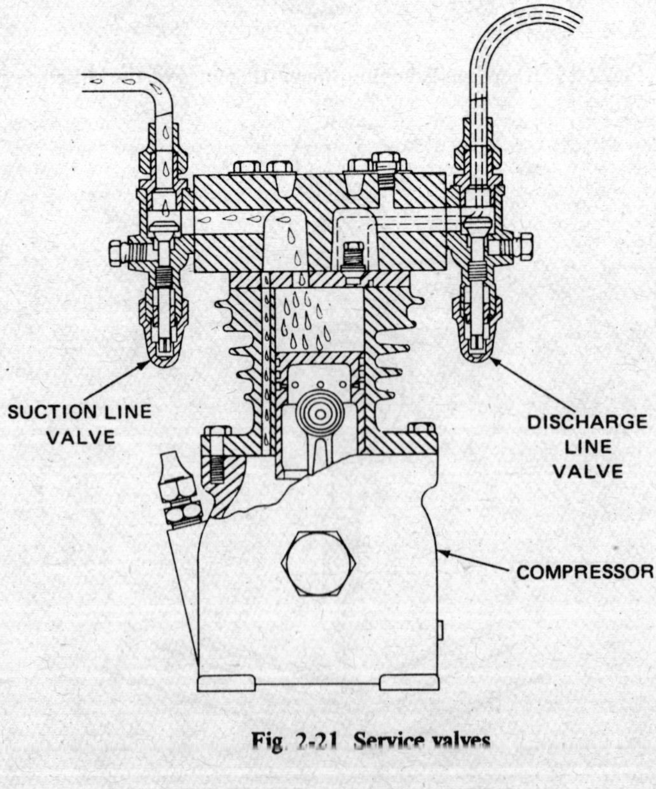

Fig. 2-21 Service valves

HIGH-PRESSURE GAUGE

COMPOUND GAUGE

Fig. 2-22 Pressure gauges

Pressure Gauges

Pressure gauges may be used to measure pressures to help determine conditions within the system. A high-pressure gauge can be used to determine pressure on the high-pressure or discharge side of the compressor. A compound gauge can be used to measure positive and negative pressures, figure 2-22.

ELECTRIC METERS

Voltmeters, ammeters, and ohmmeters are used for testing electric circuits. Voltmeters measure voltage in live circuits. Ammeters measure amperage in a circuit to determine if it is adequate. Ohmmeters measure continuity or resistance in electric circuits that are disconnected from a power source.

Always follow the manufacturer's instructions when operating these meters. The following information gives only general guidelines.

Voltmeter

A voltmeter usually has several scales for reading different levels or intensities of voltage. To use the meter, turn the scale-selector dial or knob on the face

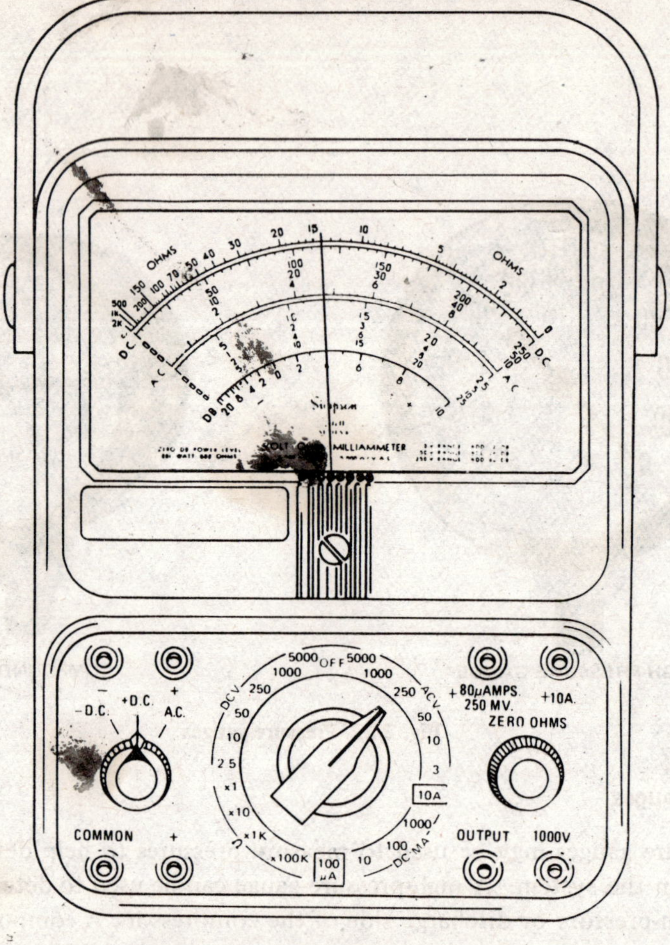

Fig. 2-23 Reading voltage on a meter. The value indicated is 115 volts.

of the meter to the correct voltage scale. For example, if the segment of the system to be tested should contain 115 volts (V), as specified in the wiring diagram, turn the selector knob to the scale that includes 115 volts. Figure 2-23 shows the selector knob at 250 volts. The meter shown in this figure is a multimeter, capable of measuring volts, amperes and ohms. In this case, the voltage is measured on the middle AC scale. Read the voltage at the point indicated. If the segment to be tested is a 24-volt, low-voltage control circuit, such as the thermostat to a gas valve, turn the selector knob to the scale that includes 24 volts. Figure 2-24 shows the selector knob at 50 volts.

CAUTION: If the voltage in the circuit to be tested is not known, begin testing with the highest scale on the voltmeter. This protects the meter from being damaged by high voltages that may be in the circuit.

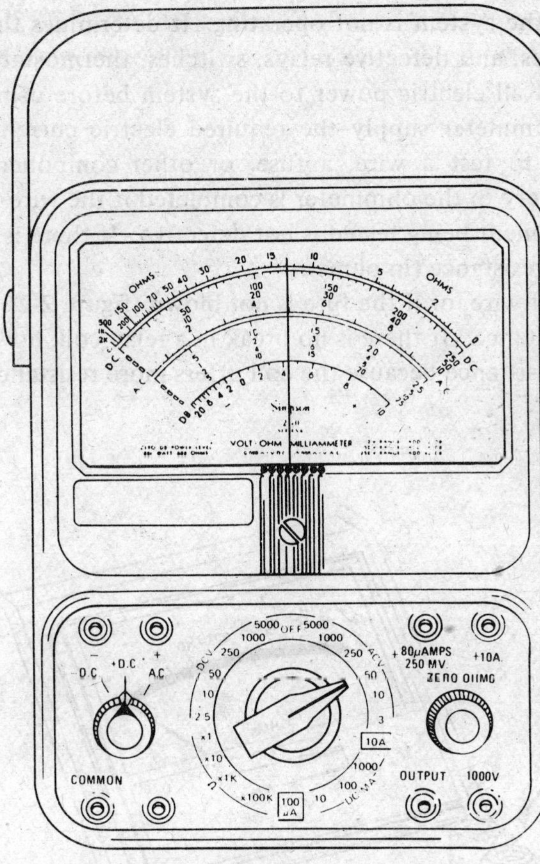

Fig. 2-24 Reading voltage on a meter. The value indicated is 24 volts.

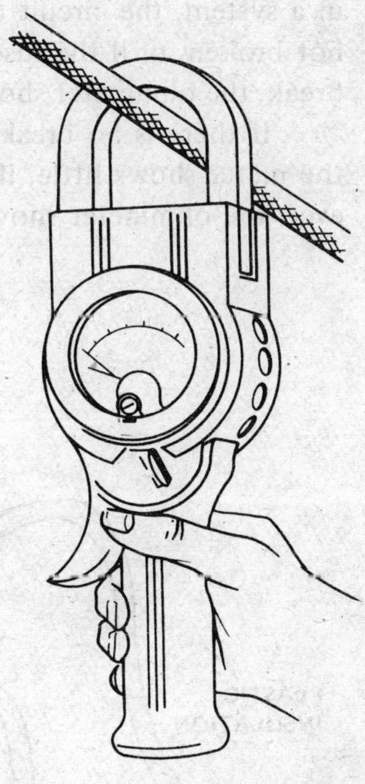

Fig. 2-25 Clamp-on ammeter

After the proper scale is selected, the meter leads are attached (clipped) to the terminals or wires of each segment or component to be tested. In the heating system described earlier, the leads are attached to circuits such as the thermostat circuit and the fan motor circuit.

Ammeter

To test the amperage in a wire, select the scale on the meter that agrees with the expected amperage. Then clamp the ammeter on the wire, as shown in figure 2-25. Be sure to test only one wire at a time. Otherwise, faulty readings will be obtained. The ammeter will indicate that the flow of electric current in the wire is adequate or inadequate compared to the amperage specified in the manufacturer's wiring diagram.

Ohmmeter

An ohmmeter is used when the system is not operating. It determines the existence of shorts or broken wires, and defective relays, switches, thermostats, fuses, and other devices. Turn off all electric power to the system before using an ohmmeter. Batteries in the ohmmeter supply the required electric current.

When an ohmmeter is used to test a wire, a fuse, or other component in a system, the circuit to the battery in the ohmmeter is completed if the wire is not broken, or if the fuse or component being tested is not defective. If there is a break, the ohmmeter shows a high resistance (in ohms).

If there is no break in a run of wire, or if the fuse is not blown (figure 2-26), the meter shows little, if any, resistance. If there is no break in a relay coil, however, the ohmmeter shows more resistance because the coil offers more resistance

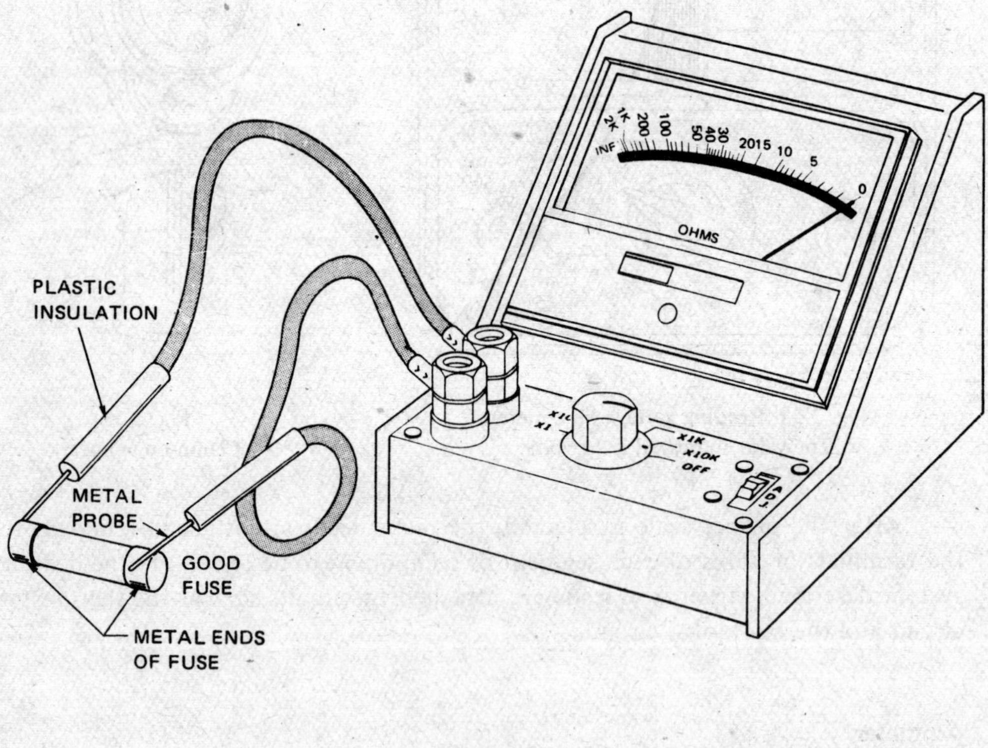

Fig. 2-26 Ohmmeter showing no resistance when connected to both ends of a fuse that is in good working order (not blown)

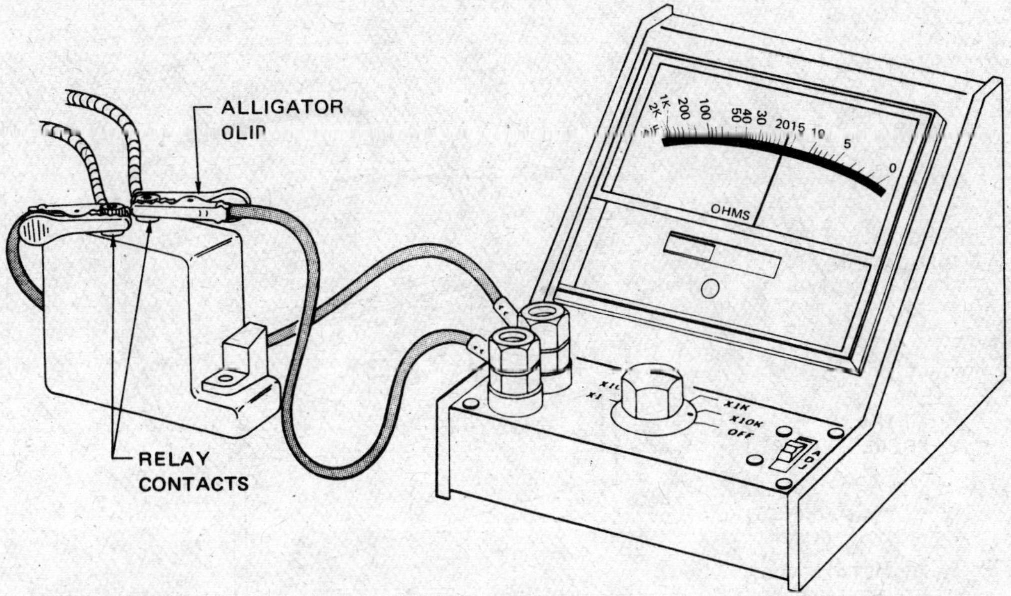

Fig. 2-27 Ohmmeter showing some resistance when connected to an operating relay coil

to the flow of electricity, and the meter needle moves somewhere in the middle of the scale, figure 2-27. If the fuse is blown or the coil is defective, the meter needle goes to the far left of the scale, indicating a maximum resistance (open break) in the segment being tested, figure 2-28. Thus, the reading on the ohmmeter depends on the segment being tested.

Additional Electrical Tools

The *test lamp* can be used to test electrical continuity in the same manner as an ohmmeter. A test lamp of suitable voltage can also be used to test the power source like a voltmeter. If there is no break in the wire tested, the lamp lights.

A flashlight and extension light are essential for use in cramped areas where permanent lighting is not adequate.

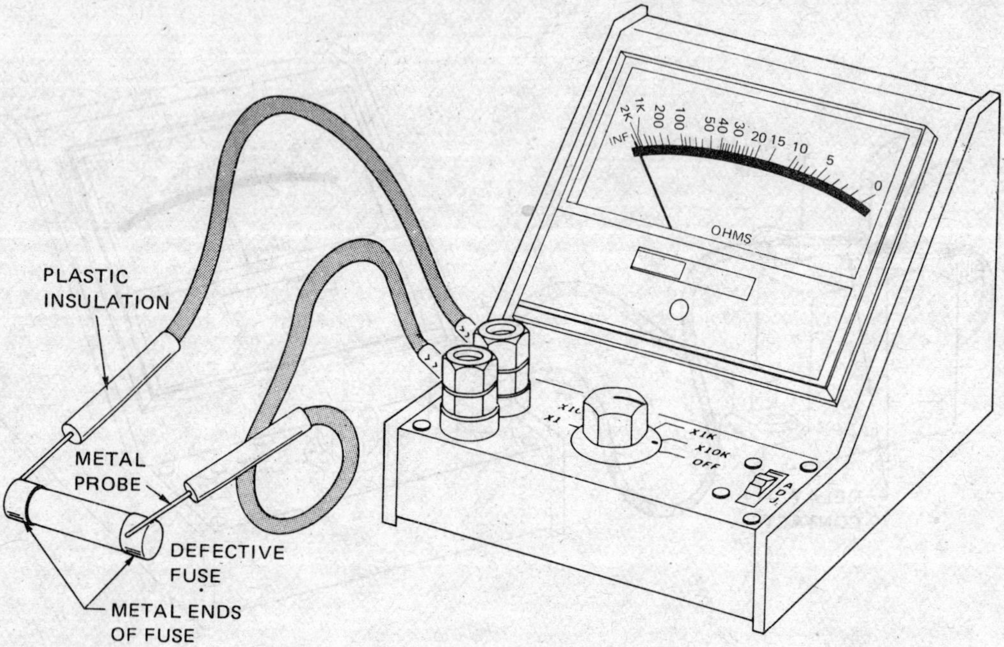

Fig. 2-28 Ohmmeter showing maximum resistance when connected to a blown fuse

AIRSIDE TOOLS

Airside tools such as thermometers, sling psychrometers, and airflow meters are used to take measurements in the airflow system. A common dry-bulb *pocket thermometer*, figure 2-29, measures discharge and return air temperature at the unit or at registers in the air-conditioned area. *Sling psychrometers*, figure 2-30, measure wet-bulb conditions or humidity in the air. *Airflow meters*, such as the velometer in figure 2-31, measure air speed.

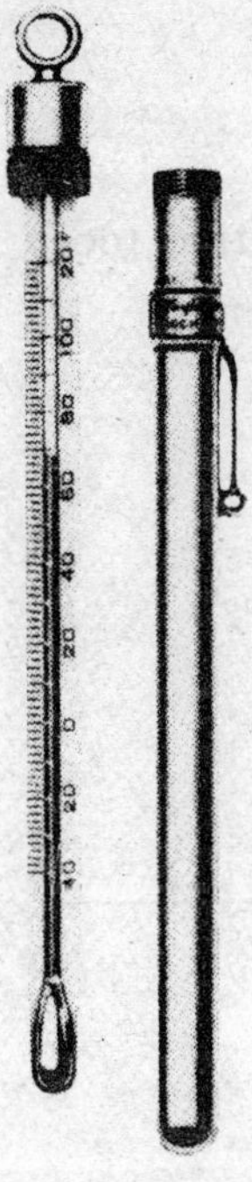

Fig. 2-29 Pocket thermometer

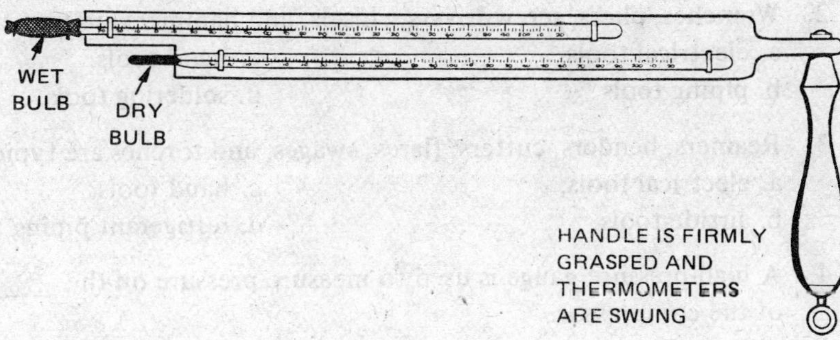

Fig. 2-30 Sling psychrometer

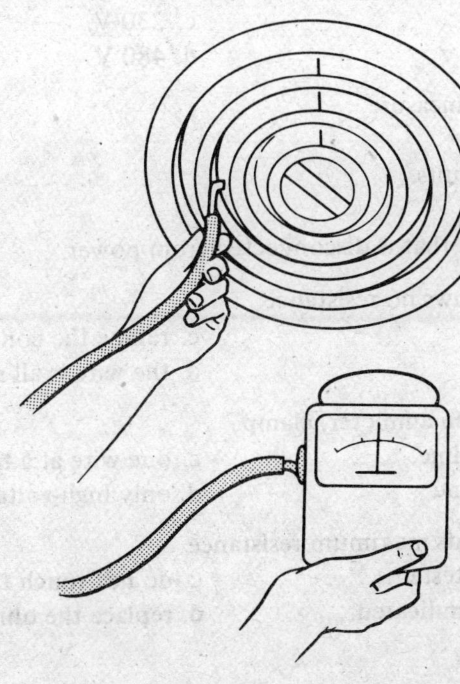

Fig. 2-31 Testing with a velometer

REVIEW

Select the best answer.

1. Which of the following is not a typical air-conditioning installation tool?
   a. Ratchet wrench
   b. Hammer
   c. Pocket thermometer
   d. Timing light

2. Wrenches, pliers, screwdrivers, chisels, and hammers are typical
   a. electrical tools.
   b. piping tools.
   c. hand tools.
   d. soldering tools.

3. Reamers, benders, cutters, flares, swages, and torches are typical
   a. electrical tools.
   b. airside tools.
   c. hand tools.
   d. refrigerant piping tools.

4. A high-pressure gauge is used to measure pressure on the ________________ of the compressor.
   a. Low side
   b. Front side
   c. Bottom
   d. High side

5. Voltmeters, ohmmeters, and ammeters are typical
   a. hand tools.
   b. electrical tools.
   c. piping tools.
   d. air-measuring meters.

6. To test a 230-volt circuit, a voltmeter scale selector should be set at
   a. 115V.
   b. 24 V.
   c. 230 V.
   d. 480 V.

7. An ohmmeter can measure
   a. 230-volt circuits.
   b. only 24-volt circuits.
   c. amperes.
   d. ohms in a circuit that is disconnected from power.

8. If an ohmmeter shows no resistance
   a. replace the wire.
   b. splice the wire.
   c. retape the connection.
   d. the wire is all right.

9. When testing with an ammeter, clamp
   a. three wires at a time.
   b. two wires at a time.
   c. one wire at a time.
   d. only high-voltage wires.

10. If an ohmmeter reads maximum resistance
    a. replace the wire tested.
    b. a high voltage is indicated.
    c. do not touch the wire.
    d. replace the ohmmeter.

3
Refrigerant Tubing, Tools, and Procedures

OBJECTIVES

After studying this unit, the student will be able to

- discuss the procedures used in forming, cutting, bending, and joining refrigerant tubing, fittings, and components.

INTRODUCTION

Before installers or service technicians can successfully complete refrigerant circuit installations, they need to know how to handle copper tubing: how to cut it, form it, bend it, and how to make joints by brazing, soldering, and flaring. To do this properly, technicians also need to know how to handle tools required for installing the piping and connecting components that are needed to form a complete refrigerant piping circuit.

Copper tubing is the most common material in a refrigerant piping circuit. It is used to connect the major components such as the evaporator coil, compressor, condenser, and expansion valve. Once connected, the tubing permits directed flow of refrigerant from one segment of the refrigerant circuit to another. Most copper tubing is available in sizes ranging from 1/8 inch through 3/4 inch.

WORKING WITH COPPER TUBING

A few basic facts about copper tubing should be helpful when working with it:

- Copper tubing, like many metals, is available in either a soft or hard form, figure 3-1. *Soft* is usually rolled in a coil and packaged. *Hard* is available in

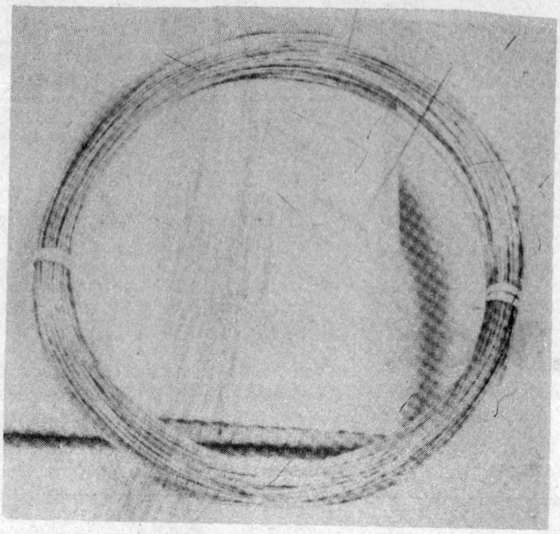

(A) Soft rolled copper tubing

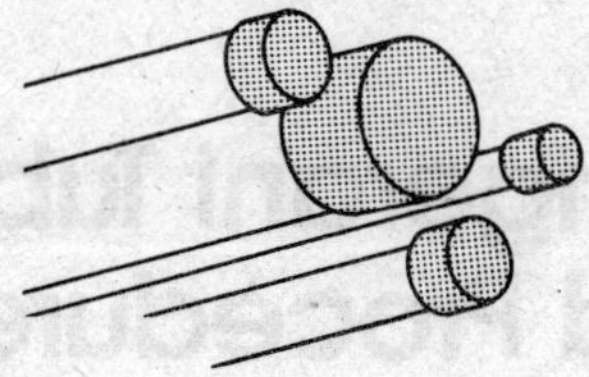

(B) Hard-drawn tubing (capped)

Fig. 3-1 Copper tubing

straight lengths and larger diameters (1/2 inch or larger). Soft copper tubing becomes hardened when it is bent or hammered. It can be softened again by first heating with a torch and then cooling with normal-temperature air.

If the packaged copper tubing is capped or pinch-sealed, internal moisture has been factory-removed. Treat the seals accordingly so moisture does not enter tubing. Moisture in a refrigerant circuit can cause the system to malfunction.

TUBE BENDING

Bending tubing by hand without the aid of a bending spring or a tube bender device causes the tube to buckle or warp out-of-round. This in turn causes it to have a short life usage and to interfere with normal refrigerant flow. In addition, if a cut is made at the damaged out-of-round area, it is extremely difficult, if not impossible, to make a reliably sealed connection at that point. Therefore, always use an internal or external bending spring or a lever-type tube bender.

Bending Springs

The internal bending spring, figure 3-2A, should just fit snugly inside the tube with a short length extending beyond the end of the tube. If the spring is totally inside the copper tube, it will be very difficult, if not impossible, to remove

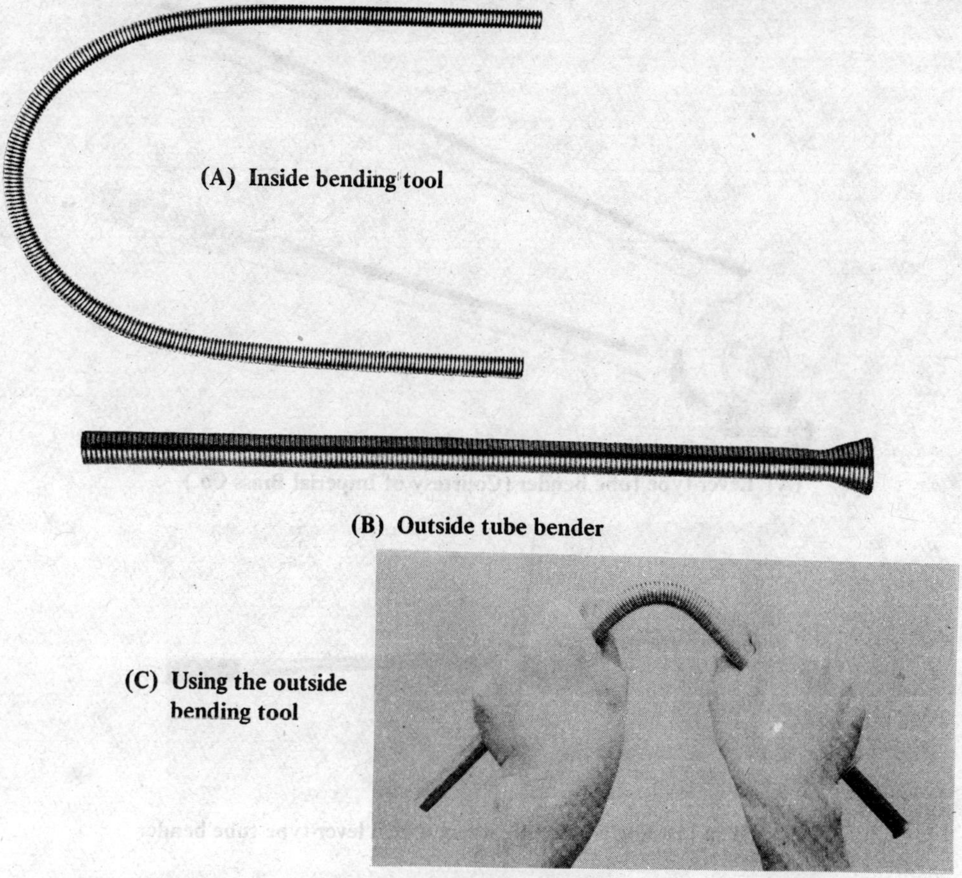

Fig. 3-2 Coil spring benders

it after the bend is made. With a proper length extending beyond the end of the tube, however, a normal twisting to cause the spring coil to contract will permit easy removal.

The external spring coil, figure 3-2B, is normally used to bend along the entire length of the tube, except at the very ends. It should fit snugly over the tubing and be moved to the area to be bent. For best results, always bend slowly and carefully. To remove the spring, use a normal twist to cause it to expand. Expand and remove the spring at the same time.

Lever-Type Bender

The use of a lever-type tube bender, figure 3-3, is almost certain to result in a normal to perfect bend. A proper size bender that fits snugly over the tubing can, in a single uninterrupted motion, bend the tubing to a 180-degree bend. A degree scale on the bender indicates the amount of bend.

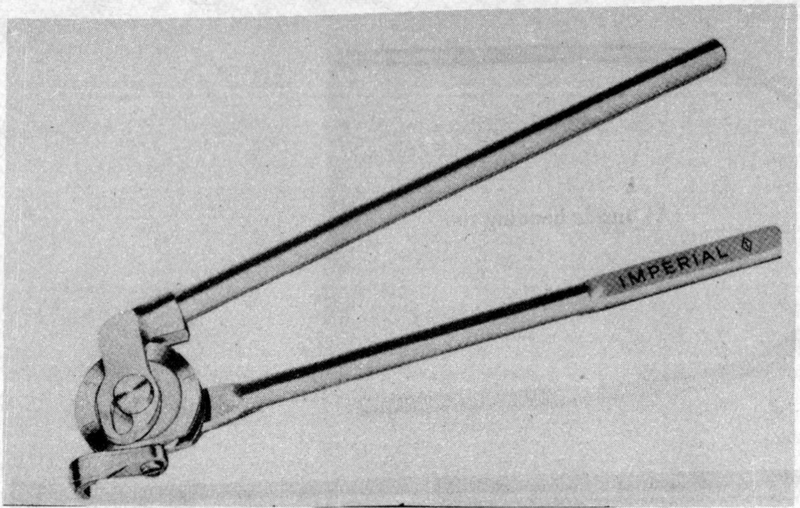

(A) Lever-type tube bender (Courtesy of Imperial Brass Co.)

(B) Step 1 in bending a piece of tubing with a lever-type tube bender

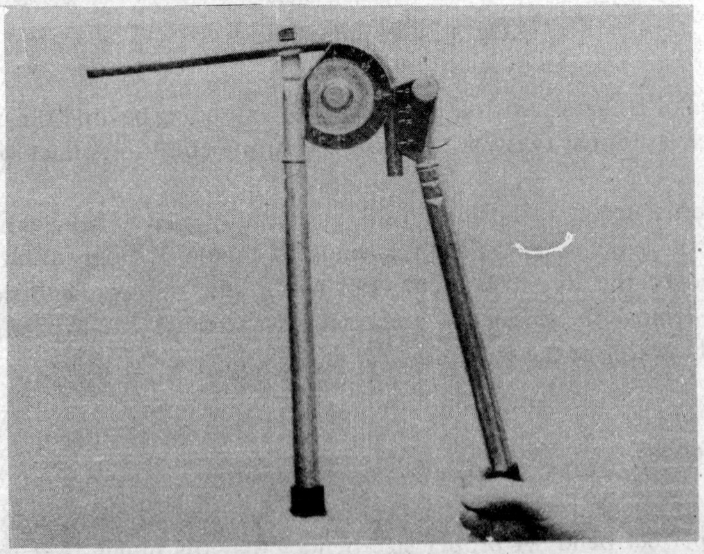

(C) Step 2 in bending a piece of tubing with a lever-type tube bender

Fig. 3-3 Lever-type tube bender

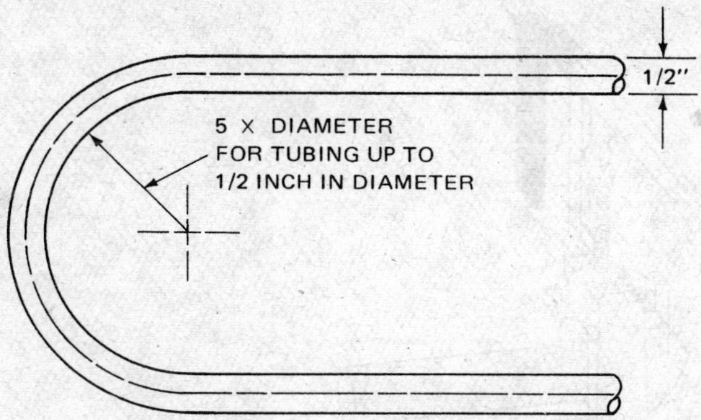

Fig. 3-4
Bending radius

Bending Guidelines

The normal bending radius for tubing up to 1/2 inch in diameter is 5 times the diameter, figure 3-4.

$$5 \times 1/2 \text{ in} = 2\ 1/2 \text{ in}$$

The normal bending radius for tubing larger than 1/2 inch in diameter is 10 times the diameter.

$$10 \times 3/4 \text{ in} = 7\ 1/2 \text{ in}$$

TUBE CUTTING

Tube Cutter

To assure an accurate "square" cut, a tube cutter is the most frequently used tool, figure 3-5. On soft copper, it permits a clean, efficient, and easy-to-make cut.

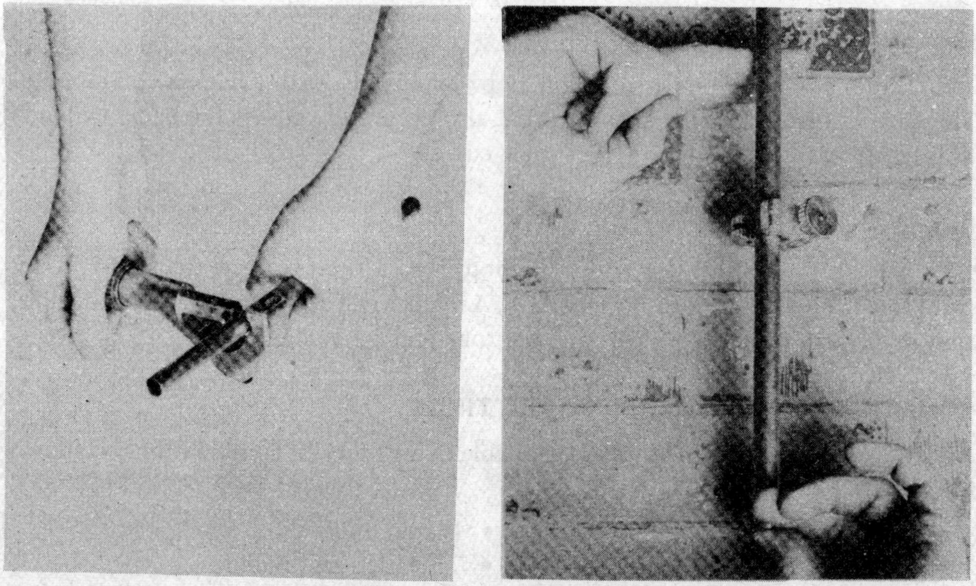

Fig. 3-5 Using a tube cutter

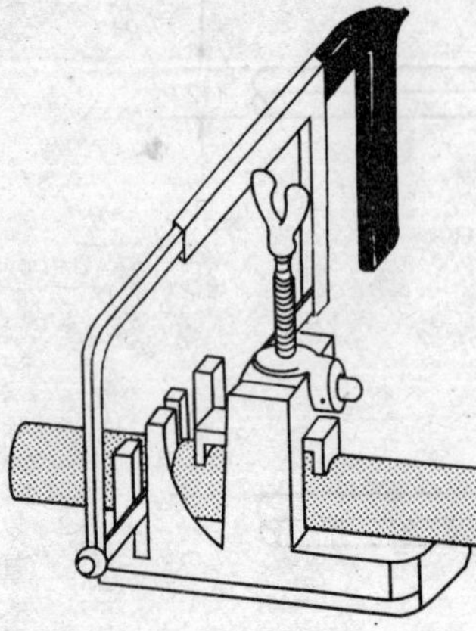

Fig. 3-6 Hacksaw and fixture

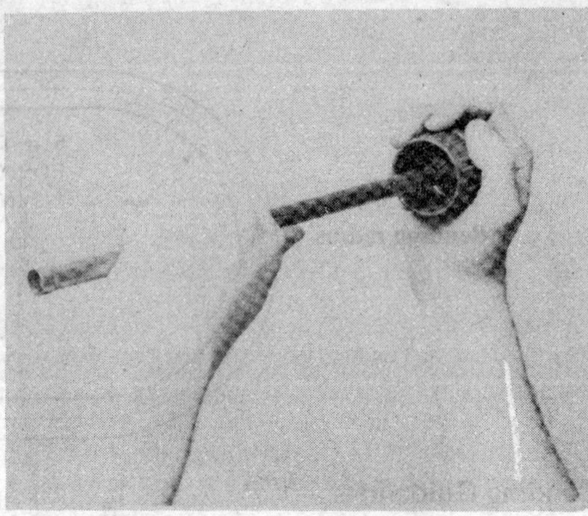

Fig. 3-7 Using a reamer

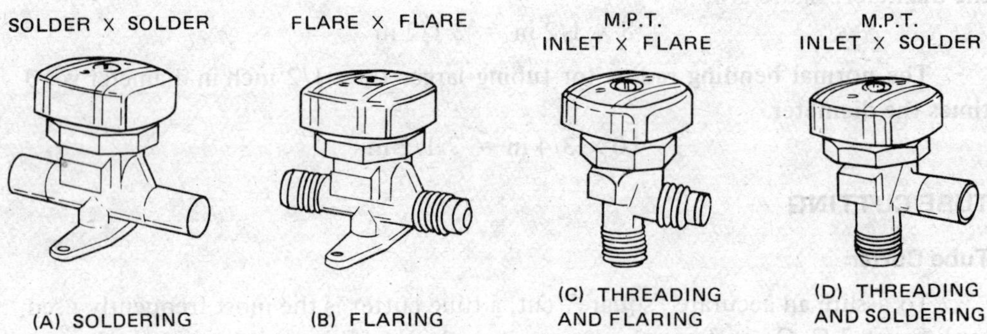

Fig. 3-8 Types of fittings

Hacksaw and Fixture

Larger size hard copper tubing may be readily cut with a hacksaw, figure 3-6. The tube is firmly held in a fixture that also accurately guides the hacksaw blade through an accurate "square" right-angle cut.

Reamers

After tubing is cut (by either method), use a tube reamer, figure 3-7, to restore the inside diameter to normal size. A reamer will also remove burrs and chips, and, in general, properly prepare the cut tube end for soldering.

COMMON REFRIGERANT PIPING FITTINGS

Refrigerant piping fittings are available in a variety of combinations, as shown in figure 3-8.

- Threaded and soldered
- Soldered only
- Flared only
- Threaded and flared

A variety of flared fittings are also available, figure 3-9.

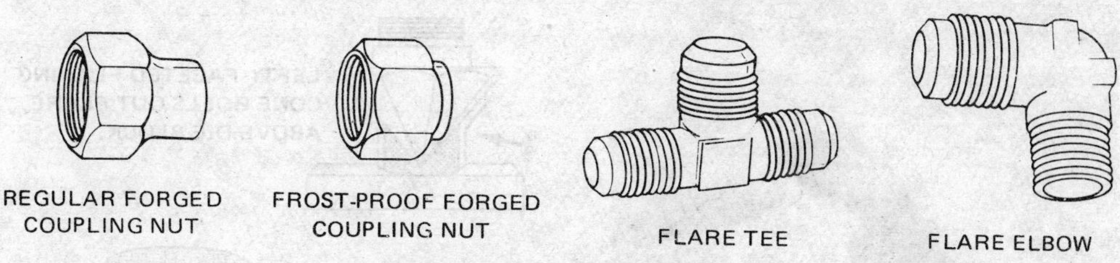

Fig. 3-9 Types of flare fittings

Types of Flaring

Single- and double-layered flares are the common types used in copper tube piping. Double-thickness flares provide greater strength. A turned (rolled) flare or punched flare are two ways of producing a double-thickness flare. The turned flare usually results in consistently accurate fitting connections. Tools commonly used to flare copper tubing are shown in figures 3-10 and 3-11.

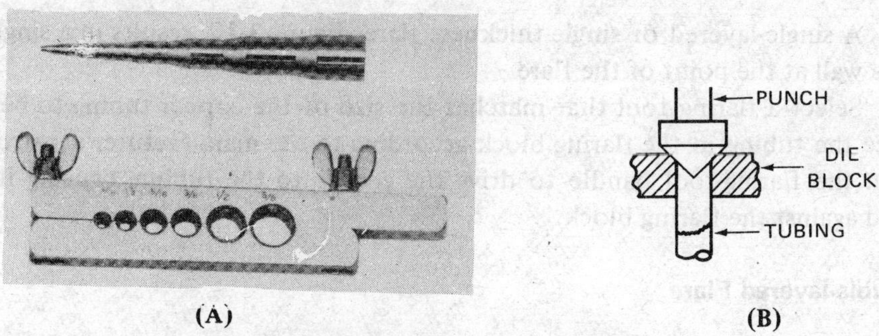

Fig 3-10 Flaring tubing punch and block

Fig. 3-11 Flaring tool

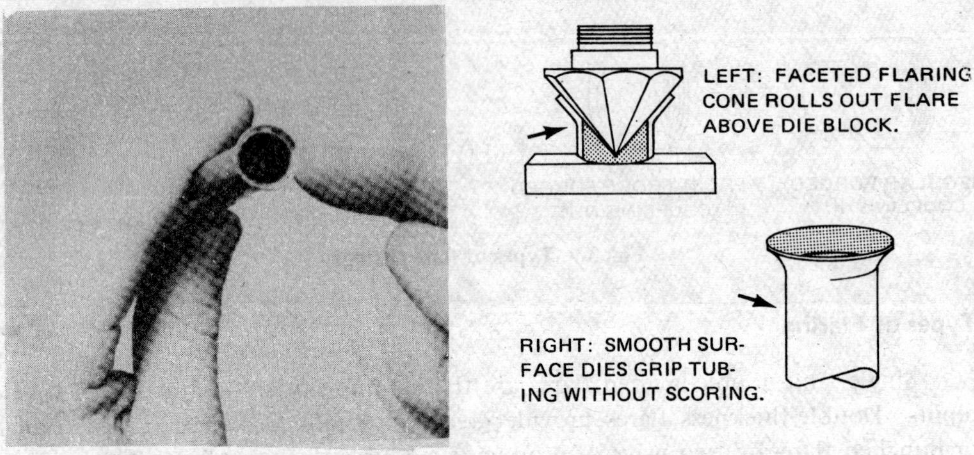

Fig. 3-12 Single-wall flare formed by the flaring tool shown in figure 3-11

Single-layer Copper Tube Flare

A single-layered or single-thickness flare, figure 3-12, results in a single-layer tube wall at the point of the flare.

Select a flaring tool that matches the size of the copper tubing to be flared. Place the tubing in the flaring block according to the manufacturer's instructions. Turn the flaring-tool handle to drive the cone into the tubing, causing it to expand against the flaring block.

Double-layered Flare

For greater strength, a double-layered tube wall at the point of flare is used, figure 3-13.

To form a double-layered flare, select the proper size flaring tool and a special adapter. Position the adapter according to the manufacturer's instructions. Turn the handle to drive the cone into the tubing, causing it to "bell-out" against the

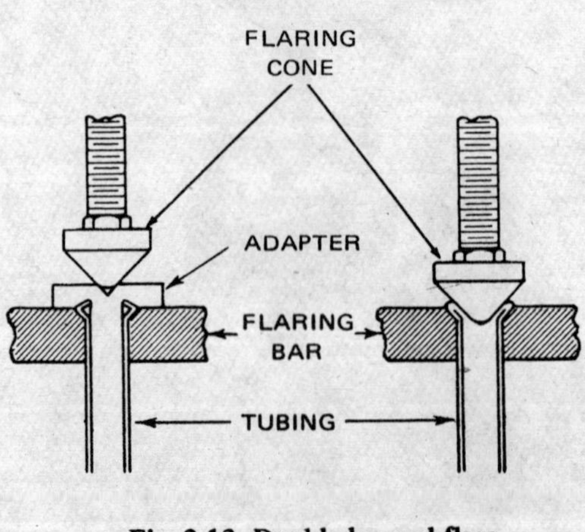

Fig. 3-13 Double-layered flare

adapter. Remove the adapter and again turn the cone into the tubing, causing it to expand into a double-walled flare.

Swaging

Swaging is a procedure by which two sections of identical size tubing can be joined permanently without using a fitting, figure 3-14.

By selecting the proper size swaging tool and following the manufacturer's instructions, the end of one section of tubing can be accurately expanded to fit snugly over the end of the second section of tubing. The "slip" joint can then be soldered using the normal procedure.

(A) Lever-type expander tool for swaging copper tubing

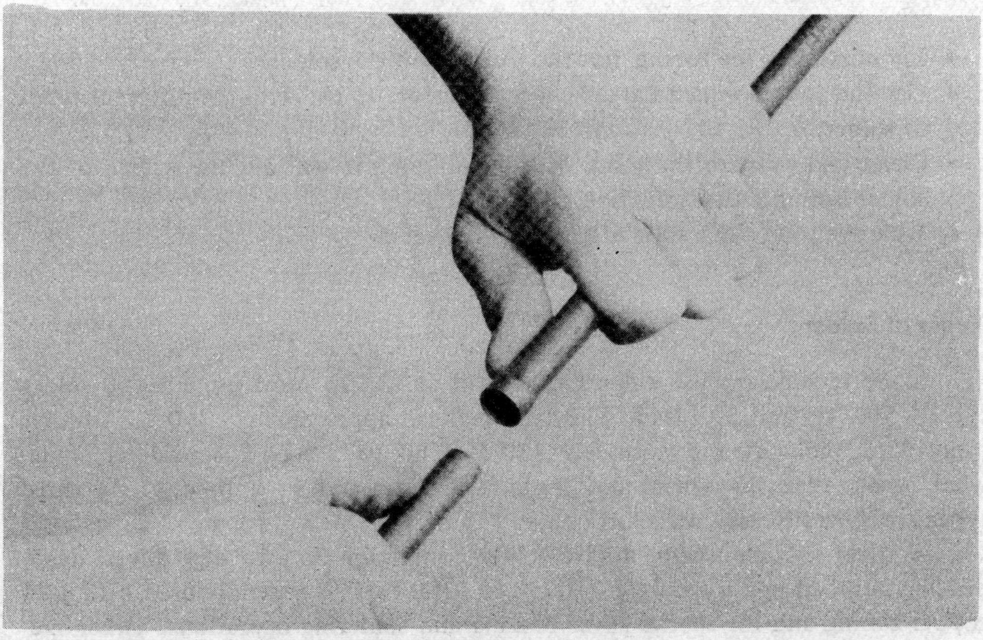

(B) Inserting tubing into the swaged end of a second piece of copper tubing

Fig. 3-14 Swaging copper tubing

TUBING SUMMARY

Soft copper tubing is available usually in rolled packages and in sizes up to 1/2 inch in diameter. Hard copper tubing is available in straight lengths in sizes about 1/2 inch in diameter.

Soft tubing becomes hardened when it is processed in such ways as bending and hammering, but it may be softened again by heating and then air cooling. Many manufacturers dehydrate and seal the tubing to keep moisture out of the refrigerant system.

All tubing should be bent with a spring bender or a special tube-bending tool. Never bend copper tube by using hands only.

Tube cutters usually produce a more perfect "square" cut. Always use a hacksaw fixture to hold the tube and guide the hacksaw blade. Reamers should be used to restore inside-tube diameter, and to remove burrs and chips.

Consistently accurate flares can be formed by using a flaring tool that turns a cone to expand the end of the copper tube. Single- or double-layered flares are possible.

A swaging tool can be used to form a joint without using a fitting.

SOLDERING

To "get the feel" of soldering takes practice. It is helpful to become familiar with some of the following techniques so that soldering will eventually, with practice, become easy.

- Let heat from the tubing, not the torch, melt the solder.
- Get the feel of which flame-heat temperature to use with the different types of solder.
- Clean and prepare the joint with extra care. It will aid the action of the solder during actual soldering.
- Wipe the joint clean soon after the solder sets.

Types of Solder

Some technicians use solders that melt at 130 F. Most use a 50/50 solder type – 50% tin and 50% lead. This type melts at approximately 360 F. Another type, 95/5, melts at approximately 450 F. This 95/5 type is considered somewhat harder than the others and contains 95% tin and 5% antimony. All three types are generally used with soft copper tubing.

A silver solder, usually a silver alloy containing 35% to 45% silver, has a melting point of approximately 1100 F to 1300 F. It is generally used with hard copper tubing and for special silver brazing and soldering operations. Other special "silver-type" solders with a melting point in the 400 F range are also on the market today.

Table 3-1 lists some common solder types.

Table 3-1 COMMON SOLDER TYPES

SOLDER	MELTING POINT	SYSTEM OPERATING TEMPERATURE	USED WITH COPPER TUBING
50/50	250 F	above 30 F	soft
50/50	360 F		soft
95/5	450 F	below 30 F	soft
silver	1100 F to 1300 F		hard

Cleaning the Joint

Cleaning the surfaces to be soldered can be done manually (mechanically) by using a wire brush, steel wool, or sandpaper. Simply use these items on the copper tubing until the inside and outside of the metals to be joined are bright and shiny. Wire brushes for cleaning the inside and the outside of copper tubing are shown in figure 3-15. The surfaces may also be cleaned chemically by using soldering flux, which eliminates any foreign material that is on the surface of the joint to be soldered.

Match Tubing and Fitting Sizes

Standard size fittings should match the tubing sizes to which they are to be soldered. The joint should fit snugly; hand tight. Too much space between the copper tube and the fitting could cause serious soldering problems. A firm, hand-tight fit is best, figure 3-16.

(A) EXTERNAL

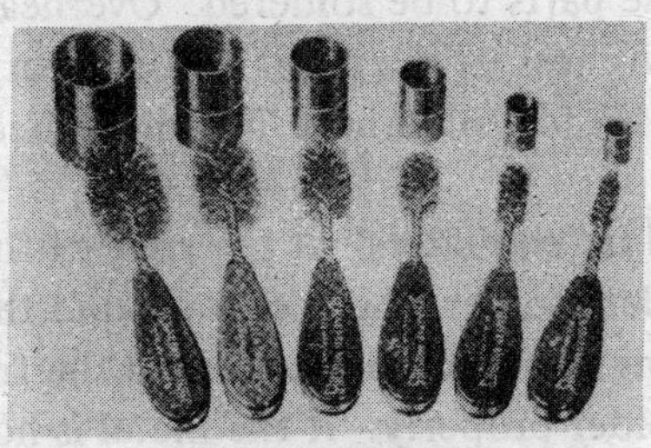

(B) INTERNAL

Fig. 3-15 Brushes (Courtesy of Schaefer Brushes)

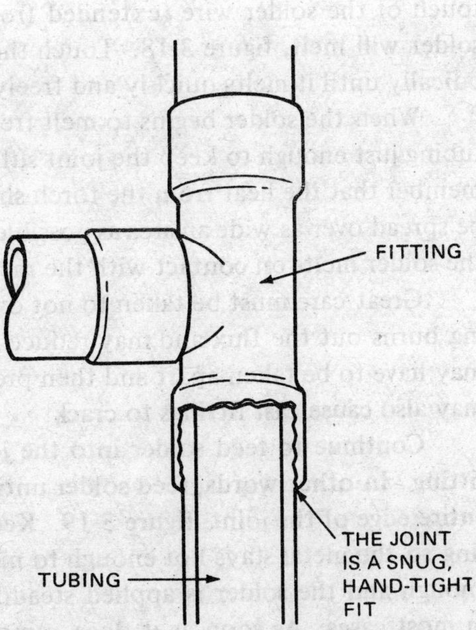

Fig. 3-16 Hand-tight fit

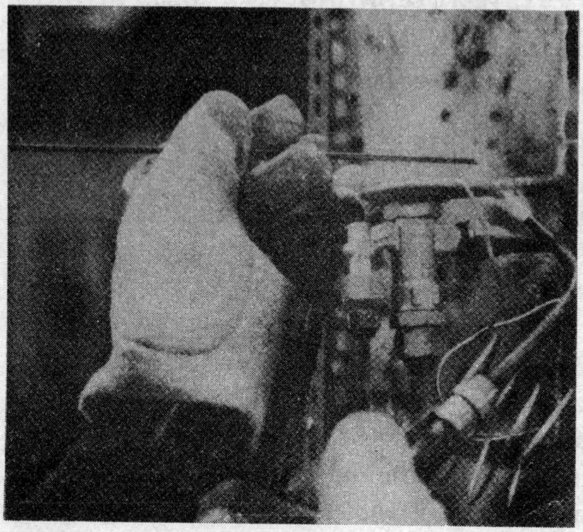

Fig. 3-17 Heat the tubing and fitting thoroughly before applying solder.

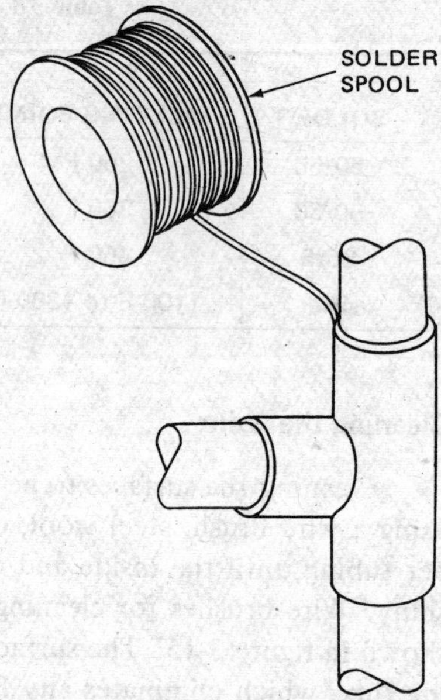

Fig. 3-18 Hot joint meets solder

Heat Joint and Apply Solder

Before applying solder, thoroughly heat the tubing and fitting on all sides, figure 3-17. Testing and experimenting will demonstrate how hot the copper must be, how red the glow of the metal, before the solder is applied. Usually a simple touch of the solder wire (extended from the solder spool) will indicate when the solder will melt, figure 3-18. Touch the solder against the hot copper tubing periodically until it melts quickly and freely and flows into the joint.

When the solder begins to melt freely, move the torch slightly away from the tubing just enough to keep the joint sufficiently hot to melt the solder freely. Remember that the heat from the torch should be directed toward the tube and must be spread over as wide an area as possible. When the correct temperature is reached, the solder melts on contact with the metal surfaces.

Great care must be taken to not overheat the parts to be soldered. Overheating burns out the flux and may reduce its effectiveness to the point that the parts may have to be taken apart and then prepared again for soldering. Too much heat may also cause cast fittings to crack.

Continue to feed solder into the joint until it "shoulders" at the edge of the fitting. In other words, feed solder until a complete solder ring occurs around the entire edge of the joint, figure 3-19. Keep working the torch flame around the fitting so the metal stays hot enough to melt the solder properly. If the metal is hot enough and the solder is applied steadily, the solder ring will form rather quickly in most cases. As soon as it does, remove the torch flame and allow the solder to

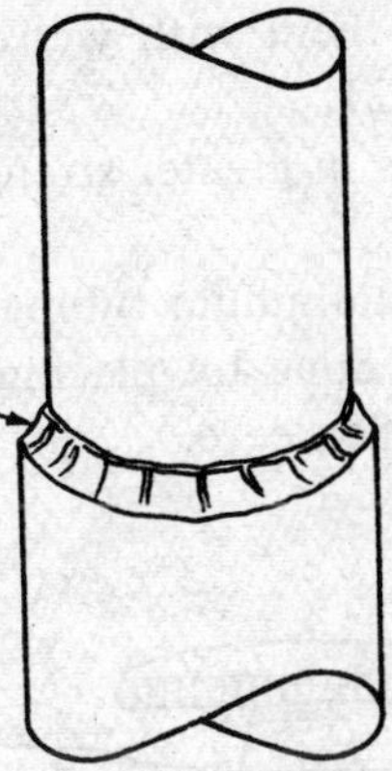

Fig. 3-19 Solder shoulder

"set." If excess solder is present, it may be wiped off the joint with a cloth. A neat-appearing solder and joint should be the result. A typical soldering torch is shown in figure 3-20.

Silver Solder

When using a silver solder, the same techniques apply regarding clean joints, flux, snug and square fit, etc., as for regular solder. The main difference is that more heat is needed to melt the silver solder. This requires a hotter torch and greater care in preventing heat damage to nearby devices such as valves and driers.

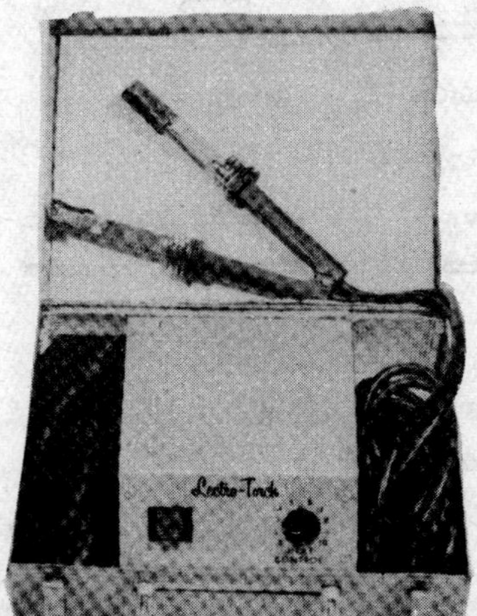

(A)

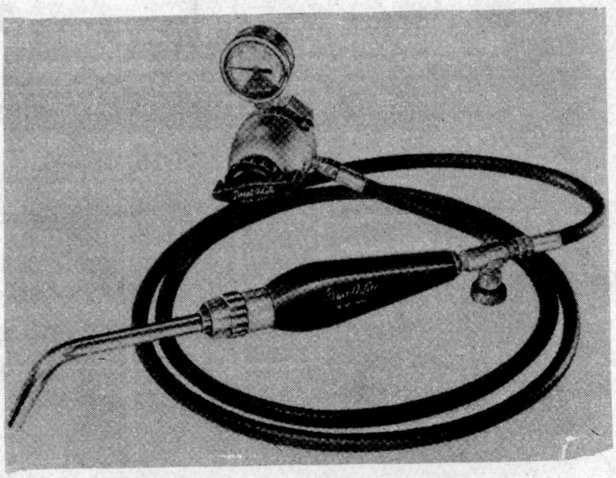

(B) (Courtesy of Linde Company)

Fig. 3-20 Torch units for soldering

To protect such devices, wrap them with wet cloths and direct flame from the torch away from them, figure 3-21.

A silver solder joint results in greater strength at the joint. See figure 3-22 for proper joint design.

When joining copper and aluminum tubing, the connection should be kept air- and moisture-free. This is done by placing a plastic sleeve over the joint, figure 3-23, and heating to seal the sleeve.

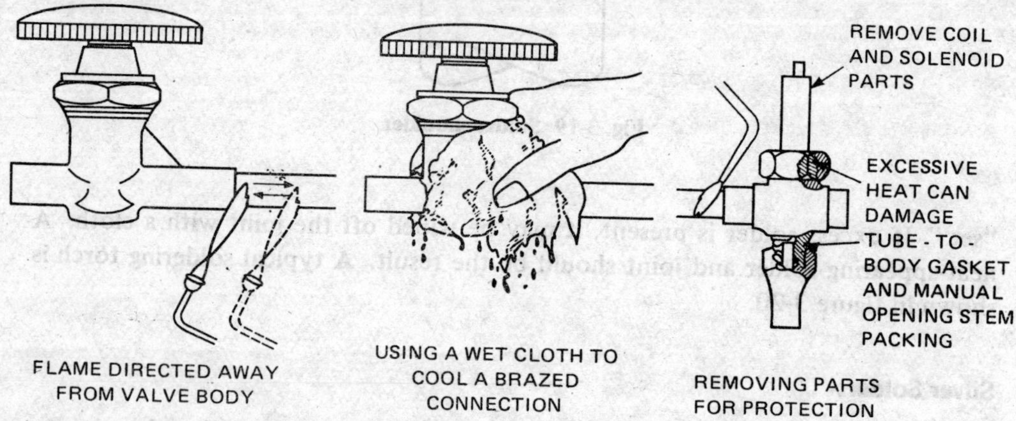

Fig. 3-21 Directing the flame to protect control devices from heat

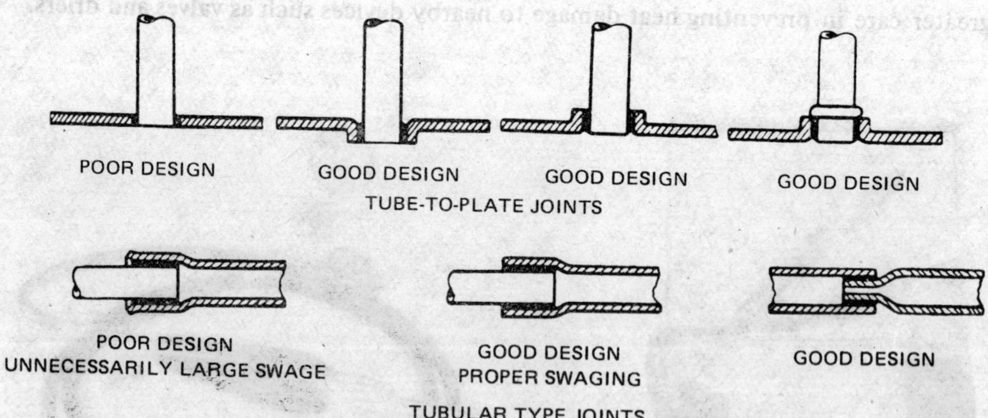

Fig. 3-22 Joints designed for silver soldering

Fig. 3-23 Copper-aluminum joints should be sealed with a plastic sleeve.

REVIEW

Select the best answer.

1. Tubing, gauges, and air-conditioning components are joined together by
   a. clamps.
   b. welding.
   c. soldering.
   d. plastics.

2. When hammered, bent, or flexed, copper tubing becomes
   a. longer.
   b. shorter.
   c. hardened.
   d. soft.

3. When heated, hardened copper tubing becomes
   a. soft.
   b. longer.
   c. shorter.
   d. egg shaped.

4. Burrs that develop on the inside of copper tubing during cutting are removed by using a
   a. saw.
   b. wet cloth.
   c. reamer.
   d. dry cloth.

5. Refrigerant tubing should be bent
   a. quickly.
   b. only in circles.
   c. slowly.
   d. only when hot.

6. A lever-type tube bender will, in one continuous operation, form bends up to
   a. 200 degrees.
   b. 260 degrees.
   c. 180 degrees.
   d. 360 degrees.

7. Tubing may be cut squarely with a
   a. torch.
   b. file.
   c. knife.
   d. tube cutter.

8. Flaring is performed accurately by using a
   a. ball peen hammer.
   b. chisel.
   c. flaring block.
   d. torch to soften the metal.

9. Tubing may be cleaned for soldering
   a. by boiling.
   b. with a torch.
   c. by using a wire brush, sandpaper, and flux.
   d. by aging in the sunlight.

10. Soft solder should be melted
    a. by the torch.
    b. only under water.
    c. by the heat of the tube being soldered
    d. by electric current.

4
Electrical Terms and Symbols, and Wire Characteristics and Sizing

OBJECTIVES

After studying this unit, the student will be able to

- define basic terms used in electricity and electric circuits.
- discuss electrical symbols and wiring diagrams.
- determine the proper wire size for an electrical installation.

BASIC ELECTRICAL TERMS

Before studying electrical symbols and diagrams, the student must have some knowledge of the common terms which make up the language used in electricity, electric circuits, and electrical wiring diagrams. The definitions of some of the most important terms follow.

Current. The terms *current* and *electric flow* mean the same thing. A strong or high current means there is a strong flow of electricity in the wire; a weak or low current means the flow of electricity is weak. High current can do much work; for example, it can operate many or large motors, or light up many bright lights. A weak or low current does less work.

Ampere is the term used to identify the amount of current or electric flow in a wire. The greater the amperage measured on an ammeter, the greater the current in the wire, and the more work that can be done.

To help explain current flow and amperes, a comparison can be made to a current of flowing water. A strong current can do much work; it can turn water wheels, or operate generators like those at Niagara Falls. Water current, or the flow of water, is similar to an electric current, or the flow of electricity. The flow of water is measured in gallons per minute, while the flow of electricity is measured in amperes.

Voltage. Voltage is the term identifying the energy that makes electric current flow in a wire. *Low voltage* means that relatively little current is flowing. *High voltage* usually means that the current flow is strong enough to seriously injure or even kill a person. Thus, it is very important to know how much voltage a wire contains at any given moment.

Resistance. Resistance is the property of any item, component, or device in an electrical system that slows down the flow of electricity by using electric energy. A light bulb, a heater, a fan motor, a compressor motor, a pump motor, a resistor device, and a partial short in the wiring all create resistance. When resistance slows down the flow of electricity, it converts the electric energy to heat or useful motion. For example, a light bulb uses electric energy to burn a small wire (filament) to create heat and light. In other words, the resistance of the light bulb converts electric energy to heat and light. An electric heater is another example of resistance in an electrical wiring system. It converts electric energy into heat. A motor converts electric energy into useful motion, resulting in such functions as turning fans, operating compressors, and operating pumps.

Ohm is the term used to identify the amount of resistance in an electrical circuit. The greater the resistance, the higher the reading on an ohmmeter.

Solenoid. A *solenoid* is a coil of wire that becomes an electric magnet when current flows through it. The electric magnet is used to do special work, such as opening or closing mechanical devices such as valves, or opening or closing electric switches. If a solenoid coil operates a valve, it is called a *solenoid valve*. If the solenoid operates a switch, it is generally called a *relay* (especially when the current flow through the solenoid coil is small). A solenoid can operate more than one switch at a time.

Switches. The main function of a switch in any electrical circuit or system is to move from one position when it is not energized to the opposite position when it is energized. The switch opens when it is energized by electric current. When it opens it either "makes" or "breaks" the flow of electricity. Switches are usually identified as *normally open* (NO) or *normally closed* (NC) when they are shown on wiring diagrams. On all diagrams, switches are shown in their nonoperating, nonenergized position (no current in the circuit).

Contactor. A contactor is a large relay designed to handle more current than a small solenoid coil relay. It operates in the same manner as a small relay but it is larger in all ways. The switches it contains are larger and the solenoid coil is larger. Therefore, the amount of current it can handle safely is also larger.

Starter. A starter is a contactor with an overload protector.

ELECTRICAL SYMBOLS

Pictures Versus Symbols

Electric wiring diagrams are pictures of all the wires, compressor motors, fan motors, small solenoid coil relays, large relay contactors, switches, lights, and other devices that may be found in most heating, cooling, and refrigeration systems.

A picture of each of these items connected by wires to form an electrical system or circuit could be shown. Such a pictorial drawing may look like the illustration in figure 4-1. By using symbols, however, the diagram is simplified, figure 4-2.

The difference is not too great when just a few items are shown in the system, or when the system is very simple. A picture diagram becomes very difficult to read, however, when all the motors, contactors, relays, switches, thermostats, lights,

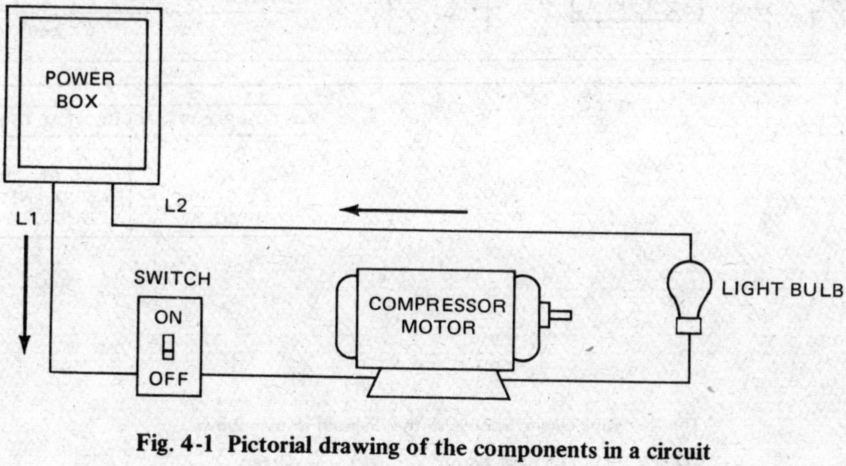

Fig. 4-1 Pictorial drawing of the components in a circuit

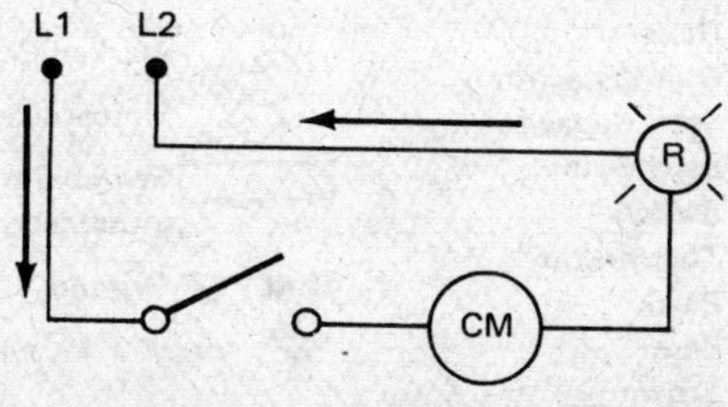

Fig. 4-2 Diagram of the circuit in figure 4-1 using symbols in place of the pictorial drawing for each device

and other components normally used in heating, cooling, or refrigeration systems are included. For this reason, wiring diagrams use symbols instead of pictures. Because there are so many different items, each diagram must have a legend to clearly identify every symbol, figure 4-3.

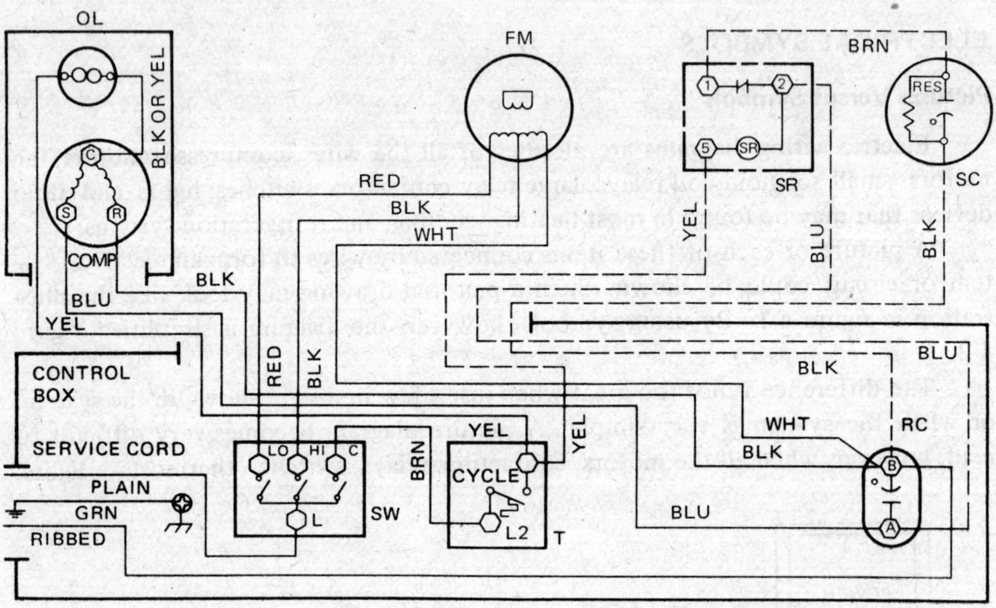

SWITCH POSITION	CONTACTS MADE
OFF	NONE
FAN	L AND HI
LO COOL	L&LO AND L&C
HI COOL	L&HI AND L&C

The following legend applies to the diagram shown above

COMP — Compressor Motor
FM — Fan Motor
Ol — Overload
Res — Resistor
RC — Run Capacitor
SC — Start Capacitor
SR — Start Relay
Sw — Switch
T — Thermostat
Blk — Black
Blu — Blue
Brn — Brown
Grn — Green
Wht — White
Yel — Yellow
⬡ Marked Connection
◇ Unmarked Connection
▭ Switch Terminal Board Connection
Gnd — Ground
⏚ or ⏚ Ground

Fig. 4-3 Typical electrical diagram showing the legend used to identify the components

Some Typical Electrical Items and Their Symbols

The following chart explains many electrical symbols used on wiring and connection diagrams.

Item	Symbol	Comment
Disconnect Switch or Fuse Box	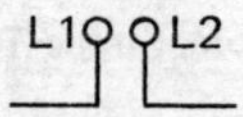	The source of electrical energy or power for the system.
Solenoid Coil	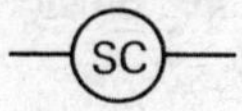	Electric magnet that opens and closes mechanical devices.
Thermostats	MERCURY 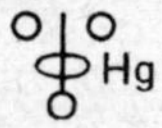	*Mercury thermostat.* A ball of mercury opens or closes the contacts inside a glass tube.
	BIMETALLIC 	*Bimetallic thermostat.* Senses temperature. A change in temperature opens or closes the thermostat contacts.
Pressurestat	 Closes on rising pressure	Measures pressure. A change in pressure opens or closes the stat.
	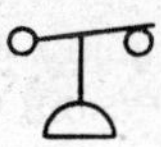 Opens on rising pressure	
Pushbutton	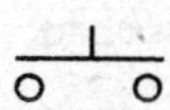 (make)	Mechanically operated by pushing with the finger to open or close.
	 (break)	
Fuse	 OR 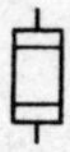 OR 	A safety device that harmlessly melts and breaks the flow of current when it temporarily becomes too high for the electric components in the system. The fuse must be sized to carry just enough current to operate the electric components in the system.
Transformer	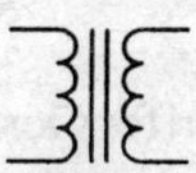 OR 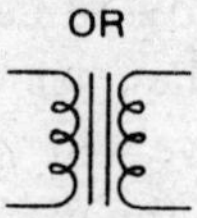	A device that changes the flow of electric current from high to low voltage. A transformer allows the simultaneous use of 220 volts for a compressor motor and 24 volts for the thermostat and other low voltage controls in the same system.

Continued

Item	Symbol	Comment
Motor	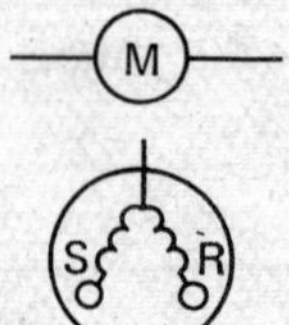	**S** identifies the motor coiling used for starting the motor. **R** identifies the motor coiling used for operating the motor after it has been started.
Switches		
SPST	 (SPST)	Single-Pole, Single-Throw.
DPST	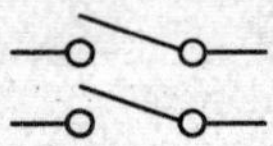 (DPST)	Double-Pole, Single-Throw.
DPDT	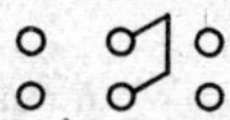 (DPDT)	Double-Pole, Double-Throw.
Contactor	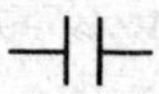 (normally open contact)	Large relay.
	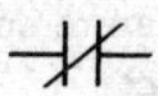 (normally closed contact)	
Overload	Magnetic  Thermal	A relay fuse or disconnect switch that protects components by tripping open or "blowing" when current flow is over the specified level.
Crankcase Heater		Keeps oil in the compressor crankcase warm when compressor is not running.
Starter		Contactor with overload.
Control Relay	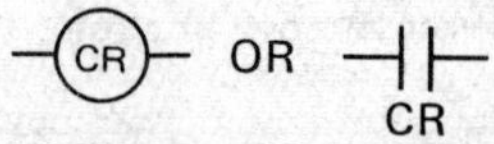	With normally open (NO) contacts.
	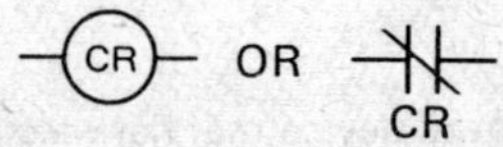	With normally closed (NC) contacts.
Solenoid Switch		
NO	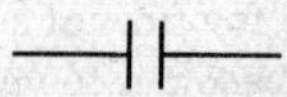	Normally open.
NC	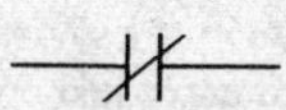	Normally closed.

CHARACTERISTICS OF ELECTRIC WIRE

Most electric wire is copper because copper offers less resistance to the flow of electric current than most other metals. The flow of current and resistance to flow of current are the key factors to consider when selecting wire, either for a new electrical circuit or for replacing old or defective wire in an existing circuit.

The flow of current can be affected by both the diameter and the length of the wire to be used. A small wire offers great resistance to the current; it tends to hold back the flow of electricity. If a tiny wire were to carry a large current, it would heat up and eventually burn, causing a break in the circuit. In addition, because a small wire holds back the flow of electricity, the supply of electrical energy would not be adequate to operate the motors and other devices in the system. Therefore, a wire that is too small can cause unnecessary but serious trouble.

A small-diameter wire may be satisfactory for a short run of 10 feet but may not be adequate for a run of 200 feet if the same supply of electric current is needed at the end of the run. The resistance to current flow at 200 feet is greater than at 10 feet, so the current at the end of the 200-foot length of wire is less. The net result is that a motor may operate properly at the end of the 10-foot wire but not operate at all at the end of a 200-foot wire. Thus, both the diameter and the length of a wire are critical to the proper operation of electrical devices.

HOW TO SELECT WIRE SIZE

One common method of sizing electric wire is based on the electric energy required to drive or operate a one-horsepower device. It is known that 746 watts are equivalent to one horsepower (hp) of electric work. But, since electric devices are not 100 percent efficient, more watts are needed. A one-horsepower motor, for example, needs 900 watts to operate properly. A discussion of the relationship between watts, the length of a wire, and wire diameter follows.

How Wire is Measured

Wire size is indicated by numbers; the smaller the wire, the larger the number. Numbers range from four zeros (#0000) to eighteen (#18). Number 18 is the smallest wire and can carry a current of only 3 amperes. Number 0000 wire is the largest and can carry approximately 195 amperes of current.

The size of a wire is determined by its cross-sectional area, figure 4-4. Since the cross-sectional areas are very small (under a fraction of an inch) they are measured in circular mils (CM). One circular mil is equal to 1/1000 of an inch. For example, the cross-sectional area of #18 wire is 1620 circular mils; #0000 wire measures 212,000 circular mils.

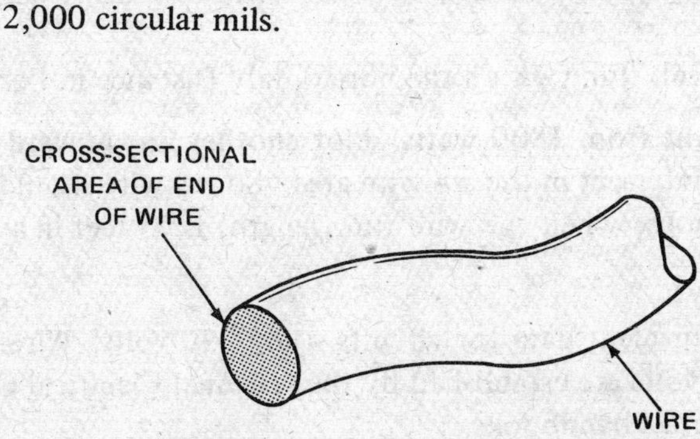

Fig. 4-4 Cross-sectional area of a wire

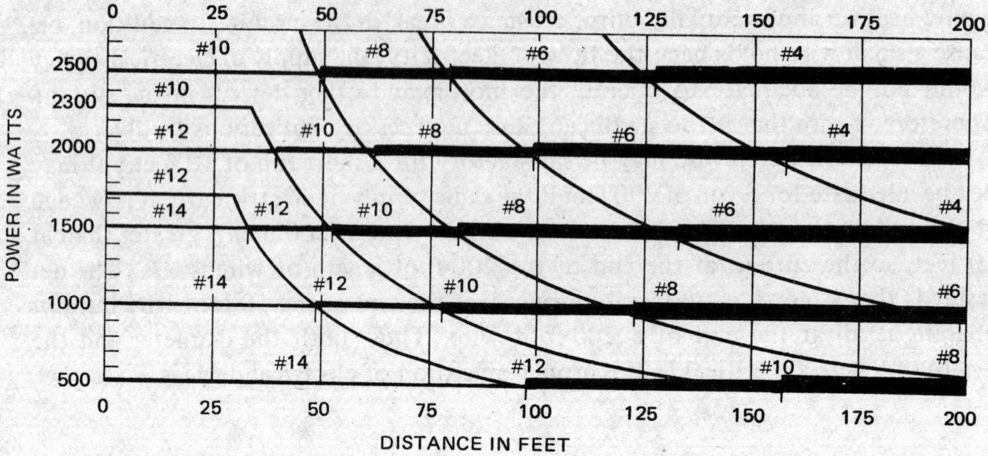

Fig. 4-5 Copper wire sizes – 115-volt circuit (2% drop)

A system of relating the power in watts to the length of a wire run for a given voltage is used to determine the wire number or size to be used. Figure 4-5 shows the wire size required for values of watts and the run distance in feet for a 115-volt circuit. For example, for a motor requiring 900 watts 50 feet from the entry (fuse) box, #14 wire is used.

Sizing Wire for Motors

Since a one-horsepower motor requires 900 watts to operate properly, a two-horsepower motor requires 1800 watts (2 × 900), and a three-horsepower motor needs 2700 watts (3 × 900).

Example. A 2-hp motor using 115 volts 75 feet from the entry box requires #8 wire. This size is determined from figure 4-5 as follows:

1. A 2-hp motor requires 1800 watts. Find 1800 on the vertical "Power in Watts" scale.
2. The distance is 75 feet. Find 75 on the horizontal "Distance in Feet" scale.
3. Plot a line to the right from 1800 watts. Plot another line upward from 75 feet. The two lines intersect in the #8 wire area. So #8 wire should be used to connect a 2-hp motor when the wire run (length) is 75 feet in a 115-volt circuit.

Figure 4-6 shows equivalent data for circuits using 230 volts. Wire sizes for other values of watts and volts are established by the National Electrical Code and are also published in electrical handbooks.

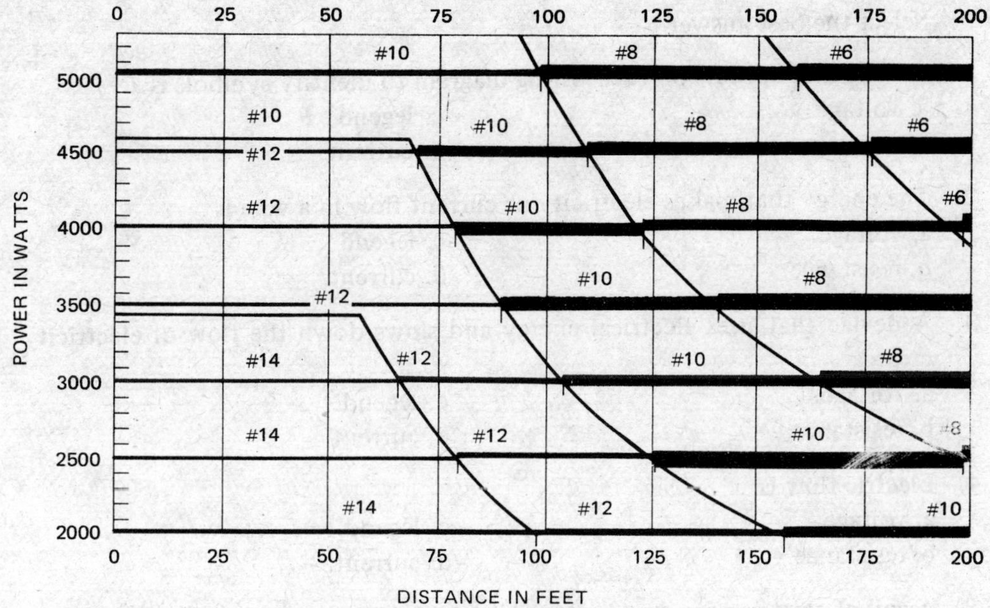

Fig. 4-6 Copper wire sizes – 230-volt circuit (2% drop)

Wire Gauge

After the required wire size is determined, the actual wire must be selected. If the wire is not marked for size, or an old wire is to be replaced, its size can be determined by removing the insulation and testing the wire end (cross section) in a gauge similar to the one shown in figure 4-7. If the wire is #10 it will just fit through the #10 opening in the gauge; if #8, it will just fit through the #8 opening, and so on.

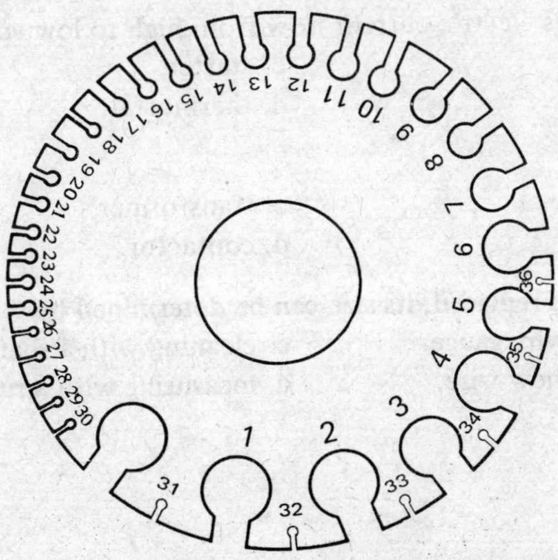

Fig. 4-7 Wire gauge

REVIEW

Select the best answer.

1. A code that appears on each wiring diagram to identify symbols is a
   a. voltage.
   b. resistance.
   c. legend.
   d. current.

2. The energy that makes electricity or current flow in a wire is
   a. voltage.
   b. resistance.
   c. legend.
   d. current.

3. A device that uses electrical energy and slows down the flow of electricity creates
   a. voltage.
   b. resistance.
   c. legend.
   d. current.

4. Electric flow is
   a. voltage.
   b. resistance.
   c. legend.
   d. current.

5. A coil of electric wire is a
   a. solenoid.
   b. switch.
   c. fuse.
   d. circuit.

6. A contactor with an overload protector is a
   a. solenoid.
   b. switch.
   c. fuse.
   d. starter.

7. A ball of mercury opens or closes contacts inside a glass tube in a
   a. pushbutton.
   b. coil.
   c. disconnect switch.
   d. mercury thermostat.

8. A device that changes electric current flow from high to low voltage is a
   a. fuse.
   b. transformer.
   c. starter.
   d. thermostat.

9. A large relay is a
   a. thermostat.
   b. fuse.
   c. transformer.
   d. contactor.

10. If an old wire is to be replaced, its size can be determined by
    a. measuring with a wire gauge.
    b. comparing with a new wire.
    c. cleaning with a sharp knife.
    d. measuring with a ruler.

5
Control Checkout and Troubleshooting

OBJECTIVES

After studying this unit, the student will be able to

- follow the general procedures for checking out air-conditioning system controls.
- troubleshoot a nonoperating system.

INTRODUCTION

Many air-conditioning control-system installers rely on experience rather than the control manufacturer's instructions when making the original installation. As a result, many control systems require a thorough checkout when making seasonal start-ups or service calls.

System controls perform essentially the same function, but they differ slightly from manufacturer to manufacturer. For this reason, the manufacturer's installation and operating instructions should be reviewed before attempting to check out or troubleshoot a system. Slightly different instructions may be needed for each installation. The general procedures discussed here can be used, provided they agree with the manufacturer's instructions.

CAUTION: Be sure to check manufacturer's instructions for electrical components before performing any electrical tests.

GENERAL CHECKOUT PROCEDURES

- Examine all wiring connections for tightness, and trace all wires to make sure the connections are made according to the instructions.
- Check the thermostat calibration and recalibrate if necessary.

- Make certain that the system starts when the thermostat calls for heating or cooling and that it stops when the thermostat is satisfied.

Thermostat Calibration (Heating and Cooling)

Use the following procedure if the thermostat requires calibration.

1. If the thermostat has a timer heater, turn the thermostat to a position that stops the operation of the system. Allow 10 minutes for the heater element to settle out.

2. If the thermostat has a mercury switch, make sure the thermostat is level.

3. To calibrate for heating:
   a. Set the system switch in the HEATING position, and the fan switch, if used, in the AUTOMATIC position.
   b. Move the set-point indicator above room temperature. The thermostat contacts should open.
   c. If the contacts remain closed, turn the heating calibration nut, figure 5-1, or heating calibration screw, figure 5-2, counter-clockwise until the contacts just open. Do not turn it beyond this point.
   d. Step back and allow the heat picked up from your hands to leave the thermostat. (This takes a few minutes.)
   e. Move the set-point indicator to room temperature.

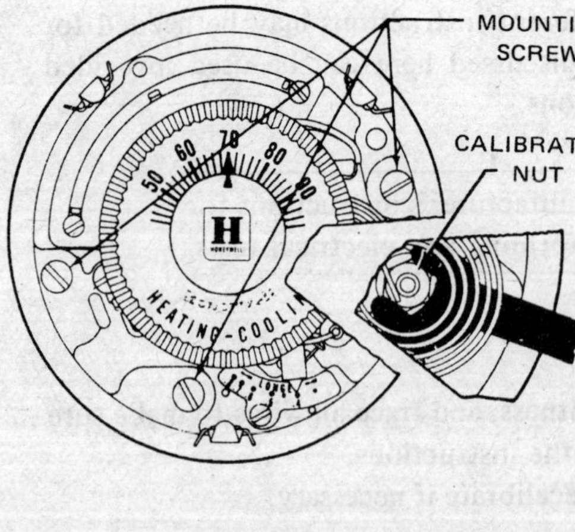

Fig. 5-1 Mercury-tube thermostat

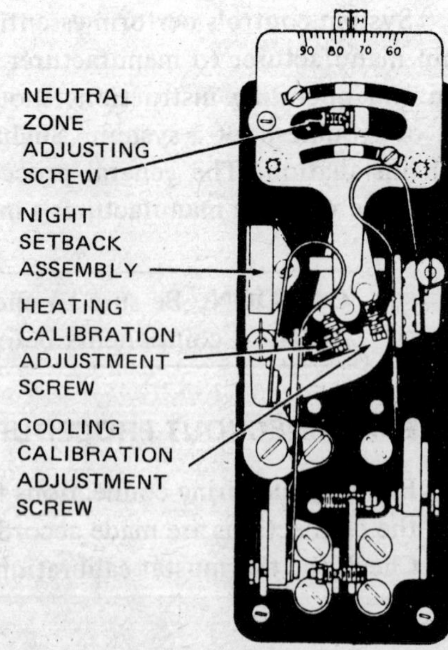

Fig. 5-2 Snap-action thermostat

f. Turn the calibration nut or screw clockwise just enough to make the contacts close. Do not turn it beyond this point.

g. Recheck the calibration. Move the set-point indicator to the lower end of the scale. Step back and wait for several minutes until the heat from your hands dissipates from the thermostat. Move the set-point indicator up-scale until the contacts just close. The contacts should close when the set-point indicator is at the same temperature as the room temperature. If they do not close, recalibrate.

4. To calibrate for cooling:

   a. Move the system switch to the COOLING position, and move the fan switch to AUTOMATIC.

   b. Move the set-point indicator to room temperature. The cooling contacts should just close.

   c. If the cooling contacts remain open, turn the cooling calibration screw or nut clockwise until the contacts just close.

   d. Turn the cooling calibration screw or nut counterclockwise until the contacts just open.

   e. Step away from the thermostat and allow the heat from your hands to dissipate.

   f. Recheck the calibration by moving the set-point indicator down-scale until the contacts just close.

   g. Slowly raise the set-point indicator up-scale until the contacts just open. The contacts should open when the set-point indicator is at the same temperature as the room temperature. If they do not open, recalibrate.

TROUBLESHOOTING THE SYSTEM

Control panels are in common use today. They simplify troubleshooting because they make it possible to check out all circuits in the system at one central point. The procedures outlined in the following paragraphs are based on a typical panel. If a centrally located panel is not used, each individual circuit must be checked at the point in the system where the component (fan motor, damper motor, thermostat, transformer) is located. Most troubleshooting can be done with suitable test light or trouble lamps and jumper wires.

When troubleshooting an electrical system that is not operating properly, the following circuits must be tested:

- Main power
- Transformer
- Relay coil
- Cooling
- Damper
- Fan
- Heating

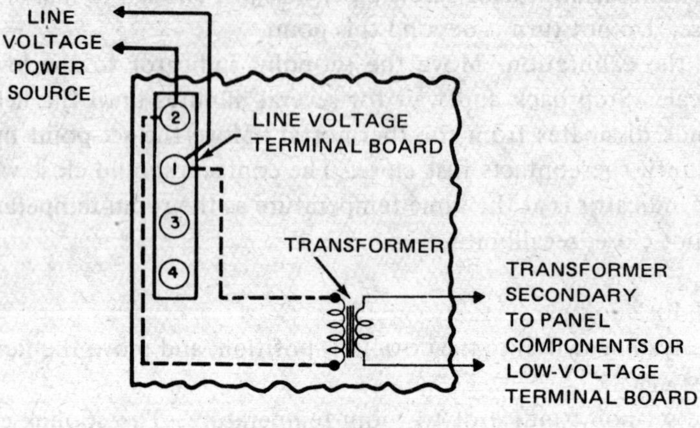

Fig. 5-3 Power source

Fig. 5-4 Checking the transformer primary

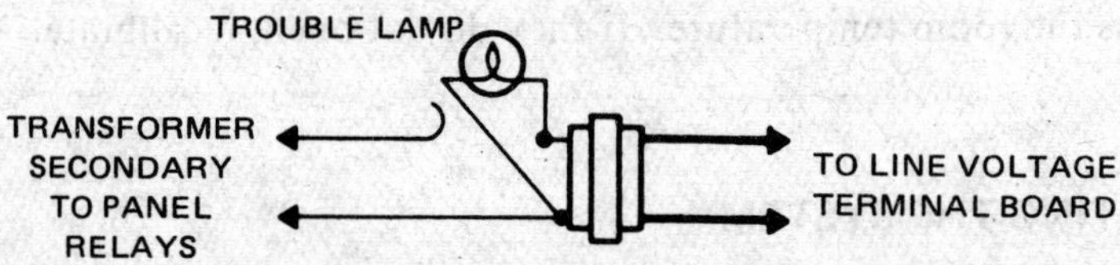

Fig. 5-5 Checking the transformer secondary

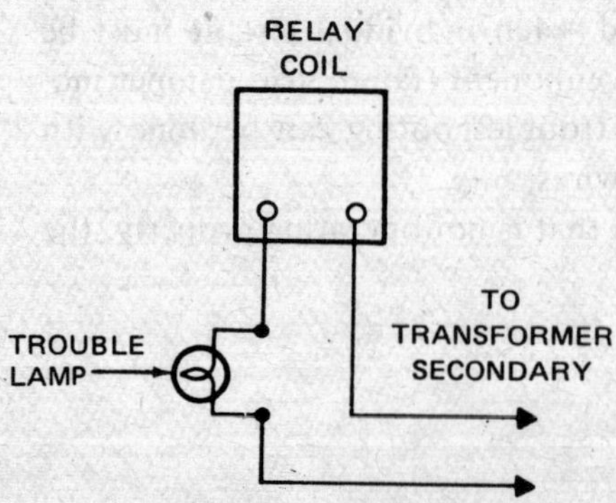

Fig. 5-6 Checking the relay

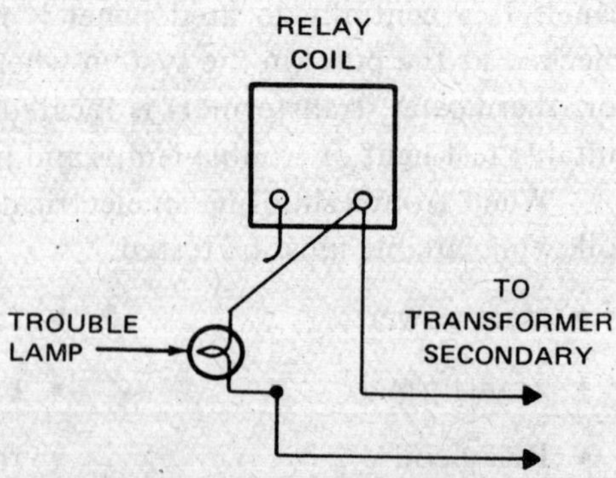

Fig. 5-7 Checking the wiring from the transformer

Power Circuit

Always start the troubleshooting procedure by testing for power ahead of the transformer.

1. Connect a test light or trouble lamp of suitable voltage (or a voltmeter) across the power terminals of the panel (terminals 1 and 2), figure 5-3. If the light goes on, the power source is operating properly.
2. If the light stays off, check for blown fuses or loose or broken connections. Correct the trouble, replace or reset the fuse, and test the power terminals again.

Transformer Circuit

1. Connect a test light or trouble lamp to the power lead on the line voltage (primary) side of the transformer, figure 5-4. If the light goes on, power to the transformer is all right.
2. If the light stays off, check the connections and the power circuit and make any necessary repairs.
3. Connect a test light across the low voltage (secondary) terminals of the transformer, figure 5-5, and disconnect one transformer lead. If the light goes on, the transformer is functioning properly.
4. If the light stays out, check and tighten the transformer line voltage connections. If the light still does not go on, replace the transformer.

Relay Coil Circuit

1. Attach a test lamp of suitable voltage to one of the relay leads as shown in figure 5-6. If the light turns on, the relay is operating correctly.
2. If the light stays off, test for power to the coil by connecting a test lamp across the circuit coming from the transformer, figure 5-7. If the light shines, replace the coil. If the light stays out, the wiring between the transformer and the relay coil is defective and must be repaired.
3. Reconnect the test light to one lead of the coil (Step 1). If the light does not burn, replace the coil. If the light goes on, the relay coil is all right.

Cooling Circuit

If any cutout devices are used, be sure they are reset before starting to troubleshoot the circuit.

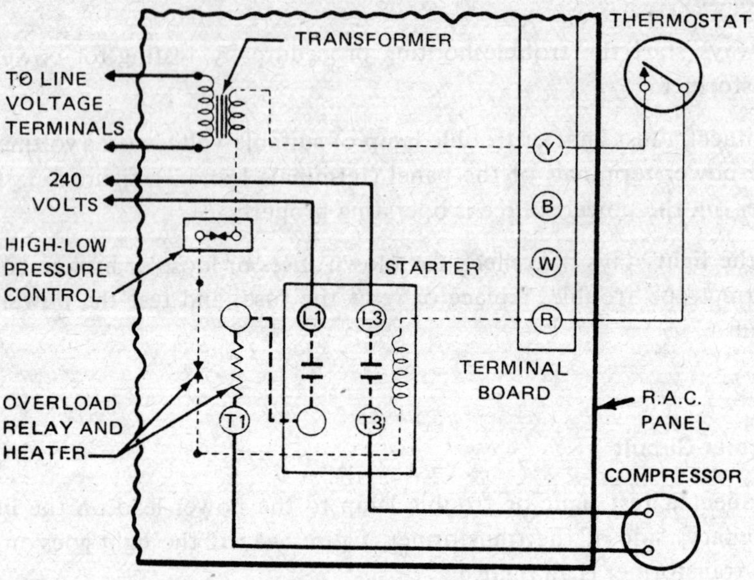

Fig. 5-8 Checking the cooling circuit

1. Attach a jumper between the thermostat leads (Y to R), figure 5-8. If the starter does not pull in, the starter coil, pressure switches, or overloads may be defective.
   a. Jumper the pressure switches and the overload. If the starter still does not operate, the pressure switches or the overload are defective and must be replaced.
   b. Replace the starter if it does not pull in after the pressure switches and overload have been replaced.

2. If the starter operates but the compressor does not operate, use line voltage caution and jumper across the starter (L1 to T1 and L3 to T3).
   a. If the compressor operates, replace the starter contacts.
   b. If the compressor does not operate, the trouble is in the cooling equipment or its wiring. Check out the wiring and each component, such as the motor and relay, and repair or replace defective wires or parts.

Damper Circuit

1. Jumper across the thermostat terminals (O to R), figure 5-9. If the damper operates, the trouble is in the control (thermostat) circuit. Make repairs or replace faulty components.

2. If the damper does not operate, the problem is in the wiring to the damper or the damper motor. Repair or replace parts as necessary.

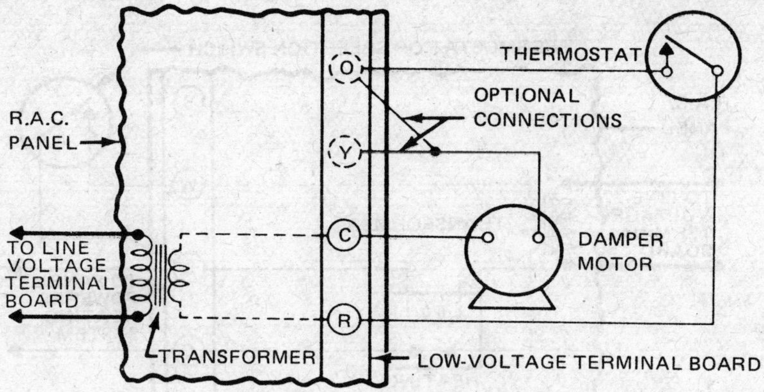

Fig. 5-9 Checking the damper

Fan Circuit

In the fan circuit shown in figure 5-10, the fan can be energized by the thermostat or through the starter.

1. Jumper across the thermostat terminals (G to R). If the fan relay does not function, replace the fan relay. If the fan starts, the control (thermostat) circuit is defective. Repair or replace it as necessary.
2. If the relay pulls in but the fan does not operate, use line voltage caution and jumper across the fan relay terminals (F to F). If the fan operates, replace the relay. If the fan does not operate, the wiring to the fan motor or the motor itself is defective. Repair or replace the motor or wiring as necessary.

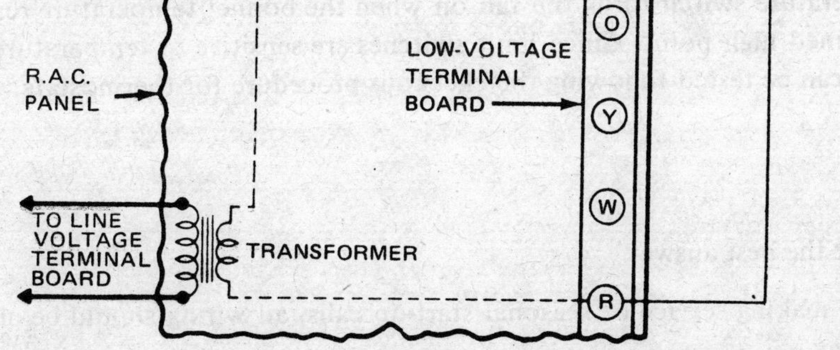

Fig. 5-10 Checking the fan

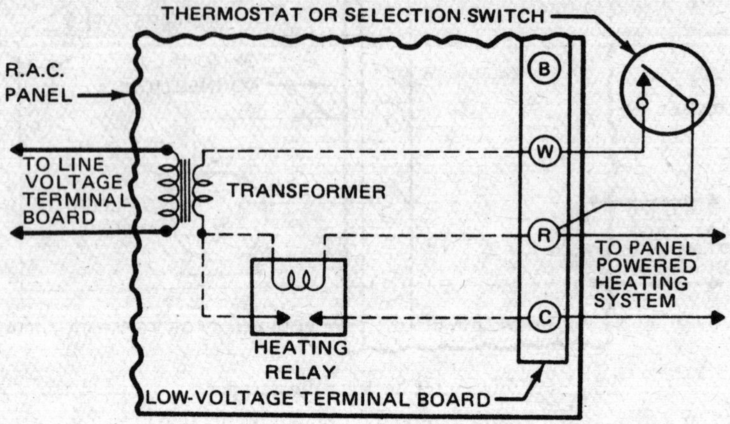

Fig. 5-11 Checking the heating circuit

Heating Circuit

1. Jumper across the thermostat terminals (W to R), figure 5-11. If the heating relay does not operate, replace the relay.
2. If the burner starts, the control (thermostat) circuit or its wiring is defective. Repair or replace the wires or components as necessary.
3. If the relay pulls in but the burner does not start, the heating components or their wiring are defective. Check out the wiring and each component; repair or replace as necessary.

Limit Switches

High-temperature and low-temperature limit switches are heat-sensing devices in the furnace bonnet. They perform like a room thermostat but are used to operate the furnace fan. When the bonnet temperature reaches a predetermined low point the low-temperature switch turns the fan off. In the same manner, the high-temperature switch turns the fan on when the bonnet temperature reaches a predetermined high point. Since limit switches are sensitive to temperature, their operation can be tested following the checkout procedure for thermostats.

REVIEW

Select the best answer.

1. When making service or seasonal start-up calls, all wiring should be checked for

   a. moisture. c. tightness.
   b. color. d. size.

2. Always begin troubleshooting ahead of the
   a. relay coil.
   b. transformer.
   c. cooling circuit.
   d. damper.

3. If the thermostat has a mercury switch make sure it is
   a. tilted.
   b. new.
   c. level.
   d. clean.

4. If a power check ahead of the transformer indicates no power
   a. stop troubleshooting.
   b. check the fuses and replace or reset them.
   c. replace the thermostat.
   d. replace the transformer.

5. Before troubleshooting the cooling circuit
   a. replace the cutout devices.
   b. reset the control devices.
   c. clean the cutouts.
   d. repaint all cutout devices.

6. If a power check ahead of the transformer indicates the power source is operating, the next circuit to be checked is the
   a. thermostat.
   b. damper.
   c. transformer.
   d. relay coil.

7. If a voltage check indicates a defective transformer circuit, check the voltage again after first
   a. tightening the line voltage power connections.
   b. replacing the thermostat.
   c. replacing the relay.
   d. cleaning the transformer surface.

8. If the transformer is still defective after taking the corrective action indicated in question 7
   a. replace the relay.
   b. replace the transformer.
   c. replace the thermostat.
   d. replace all of these.

9. If the fan relay does not operate, replace the
   a. thermostat.
   b. damper.
   c. fan relay.
   d. starter coil.

10. Limit switches control
    a. the length of time a burner operates.
    b. temperature in the bonnet.
    c. space limitations.
    d. the height of the flame.

6
Preparing to Install the Air-conditioning Unit

OBJECTIVES

After studying this unit, the student will be able to

- inspect a new air-conditioning unit and the desired installation site.
- describe precast and built-up base pads and the proper construction of each.

BEFORE INSTALLATION BEGINS

Inspect the Unit

Uncrate the unit at the job site and carefully inspect it for shipping or handling damage and for missing parts. If necessary, immediately file a detailed claim with the shipping company and/or the manufacturer. Make a full report of damage before attempting to install the unit.

Inspect the Installation Site

If your inspection indicates that the unit is ready to install, move it to the desired installation site. Locate the manufacturer's installation instructions, and study them carefully. Evaluate the site by comparing it with the instructions and by answering the following questions:

- Is there sufficient space available for making electrical connections, and is this space accessible?
- How much wire is required?

Fig. 6-1 Select an installation site which allows sufficient space for servicing.

- Is there adequate space for servicing the unit, and for removing or opening the access doors? See figure 6-1.
- Is there enough space available for making piping and duct connections?
- How much piping and ductwork is required?
- Is the location of the unit suitable? Will rainwater or ice from the roof damage it in that location? If so, select a safer site, figure 6-2.
- Will the unit have enough free airflow for efficient operation, as indicated in the manufacturer's instructions? If not, move the unit or remove the obstructions to the airflow.

Perform any additional site-review or related actions indicated by the manufacturer's instructions.

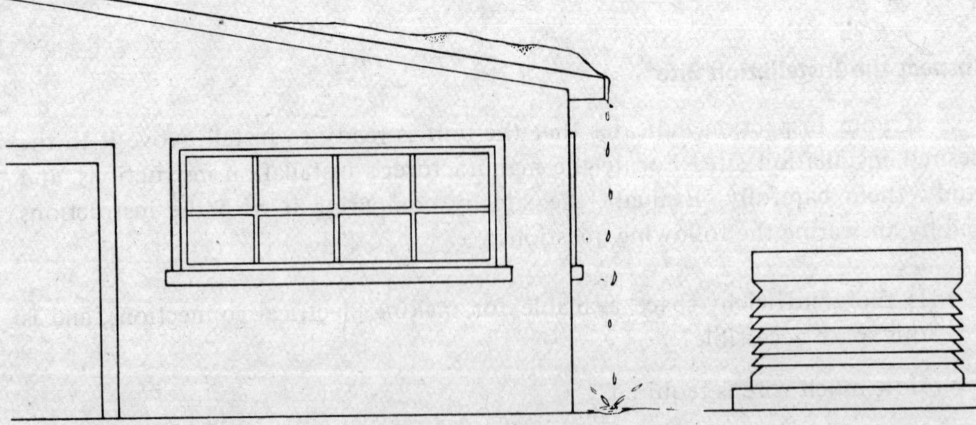

Fig. 6-2 To prevent rain or ice damage, the unit must be located away from the eaves of nearby buildings.

Check the Nameplate Data

Before the unit can be installed, the nameplate, figure 6-3, must be studied. Nameplates contain vital power and assembly data.

- Check the electric power supply available and compare it with the power rating listed on the nameplate. They must be identical.
- Check the type of gas available at the site and compare it with the type of gas listed on the nameplate. They must agree.
- Check both the cooling and the heating capacities required for the job and compare these values with the capacities listed on the nameplate. They must agree.

If the power supply, gas type, or the capacities required for the job do not agree with the nameplate listings, the unit should not be installed until the differences are resolved.

BASE PADS FOR OUTDOOR UNITS

Each manufacturer establishes its own specific requirements for a suitable base on which to set an outdoor unit. Follow the manufacturer's guidelines to select the proper type of base and installation method.

Precast Cement Base

A precast cement base pad is commonly recommended. The thickness varies depending on the size and weight of the unit. The pad should be placed on level ground which is covered with several inches of gravel.

The shape of the base pad should conform to the shape of the outdoor unit – circular for a round unit, and rectangular for a rectangular unit. The size of the pad should be large enough to extend several inches beyond the dimensions of the unit. Gravel aprons usually extend, 1 to 2 feet beyond the pad to allow for drainage and unobstructed airflow, figure 6-4. Follow the manufacturer's recommendations.

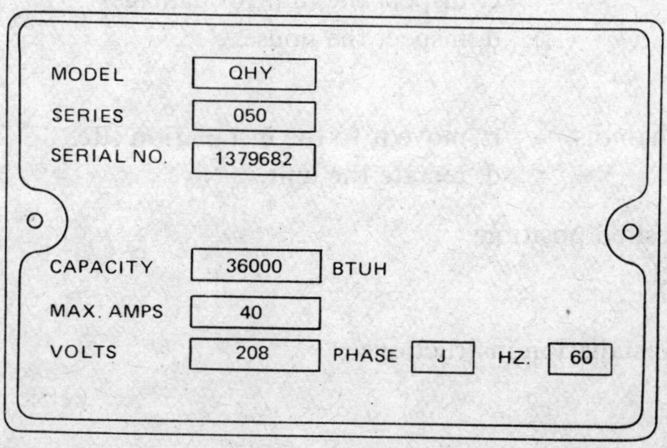

Sample of a typical nameplate

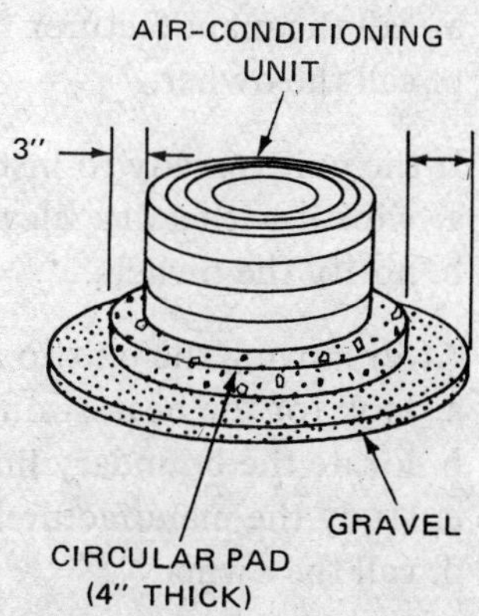

Fig. 6-4 Unit installed on a gravel apron

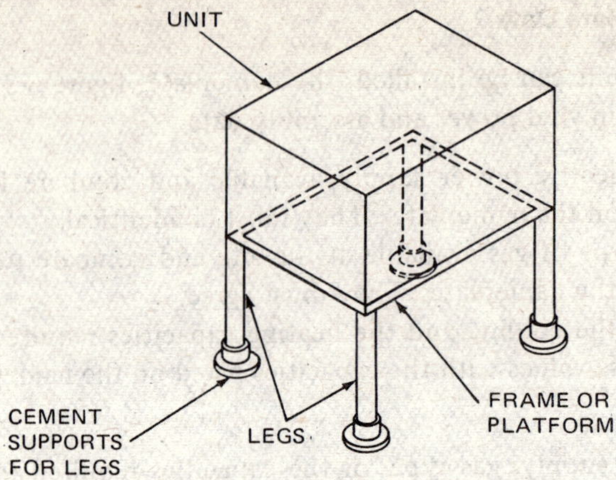

Fig. 6-5 Off-the-ground residential pad

Wooden- or Iron-post Base Pads

In snow country or in any location where conditions require the unit to be installed above the ground, wooden or iron posts may be used. For a more permanent installation, the posts should be set in concrete. Follow the normal construction practices for the climate and surface conditions at the job location.

A suitable frame or platform should be attached (preferably bolted) to the posts, figure 6-5. The frame or platform must be large enough and strong enough to hold the weight of the unit. Several manufacturers offer metal-frame bases designed specifically for units to be installed off the ground.

REVIEW

Select the best answer.

1. Before installation begins
   a. call the manufacturer.
   b. call the owner.
   c. inspect the unit for damage.
   d. inspect the house.

2. If the unit is ready to install
   a. wipe it with a dry, clean chamois.
   b. notify the owners.
   c. move it to the installation site.
   d. recrate the unit.

3. After moving the unit to a desired position
   a. look for the nearest outlet.
   b. locate the boundary lines.
   c. study the manufacturer's installation instructions.
   d. call the owner.

4. Check the unit nameplate data and compare it with
   a. the owner's bill of sale.
   b. a competitive unit.
   c. the previous installation.
   d. the power supply, type of gas, and capacities required.

5. If the power supply, gas type, or capacities required for the job do not agree with the nameplate data
   a. check with a neighbor.
   b. do not proceed until the differences are resolved.
   c. proceed with the installation and check later.
   d. proceed with the installation and write a letter to the manufacturer.

6. Base pads for outdoor units are usually made of
   a. hard rubber.
   b. foam.
   c. precast concrete.
   d. hollow plastic.

7. Base pads are usually placed on
   a. a firm slope to the right.
   b. a gradual slope to the left.
   c. a straw bed.
   d. several inches of gravel on a level area.

8. Gravel aprons usually extend beyond the unit pad at least
   a. one to two feet.
   b. six feet.
   c. ten feet.
   d. seven inches.

9. A wooden- or iron-post base is used
   a. in hilly terrain.
   b. in snow country.
   c. in the backyard only.
   d. to satisfy the neighbors.

10. For a more permanent installation, posts should be
    a. painted rust red.
    b. set in sand.
    c. set in concrete.
    d. wound with metal tape.

7
Installing Room Air Conditioners

OBJECTIVE

After studying this unit, the student will be able to

- install and service a typical room air conditioner.

INTRODUCTION

Several types of room air conditioners are available. They are designed to provide cooling only, cooling and heating using strip headers, or cooling and heating using heat pumps. The methods of installation also vary. There are through-the-wall units, figure 7-1; window units, figure 7-2; and casement window units, figure 7-3.

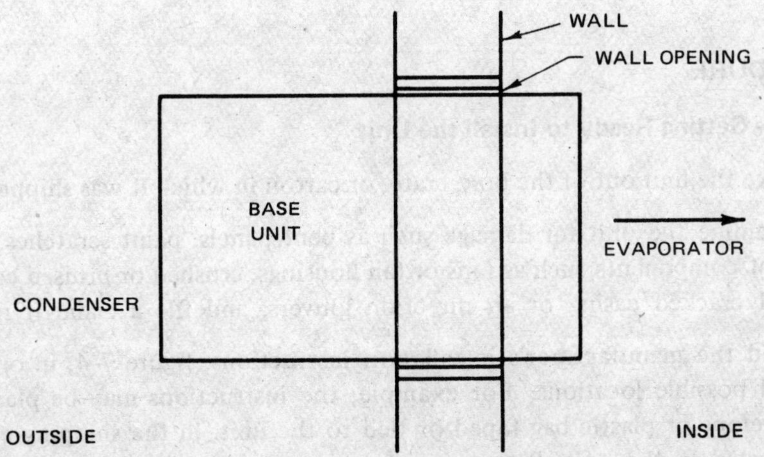

Fig. 7-1 Base unit projecting through wall

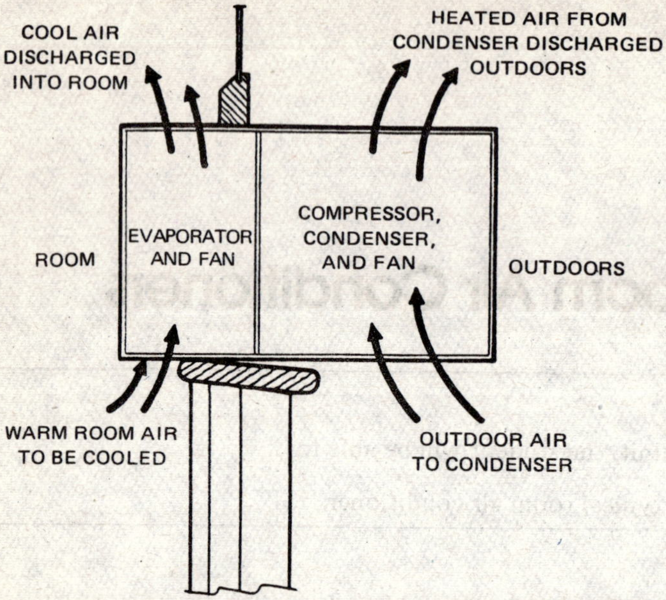

Fig. 7-2 The window-type room cooling unit

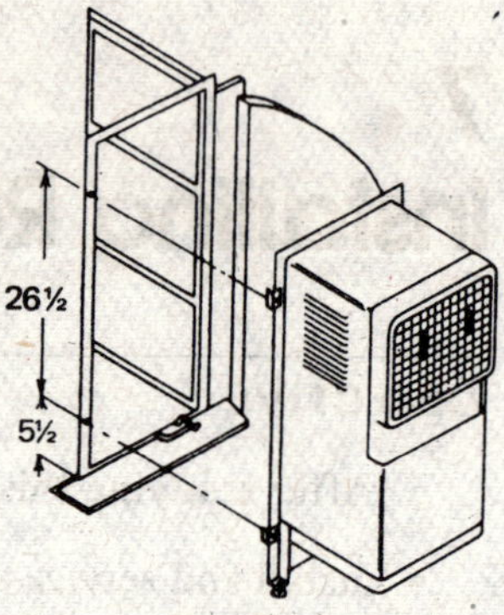

Fig. 7-3 Casement window unit

Although installation, start-up, and service details often differ from manufacturer to manufacturer, the general procedures outlined in this unit apply to most brands. Specific details are covered in the manufacturer's instructions available with each unit. The procedures which follow apply to a cooling-only room air conditioner. Refer to Units 18 and 19 of the text *Principles of Air Conditioning*, 3rd edition (Lang, Delmar Publishers), for additional room air-conditioner installation and service details.

PROCEDURE

Step 1 – Getting Ready to Install the Unit

1. Take the unit out of the box, crate, or carton in which it was shipped.
2. Examine the unit for damage such as bent panels, paint scratches, loose or bent components such as fans or fan housings, crushed or bruised coil grilles, and cracked casing or air direction louvers, and file a claim if necessary.
3. Find the manufacturer's installation instructions, figure 7-4, in one of several possible locations. For example, the instructions may be placed in an envelope or plastic bag taped or tied to the unit, in the shipping carton, or inserted in the unit. Some manufacturers insert the instructions in the filter area behind the front grille.
4. Depending upon manufacturer's policy and shipping techniques, a list of the items packed in the shipping carton may be included in the front section of

Room Air Conditioners

INSTALLATION INSTRUCTIONS

BEFORE INSTALLATION

Check available power supply and resolve any wiring problems. Refer to Electrical Data.

Test run unit.

Determine location for unit. Make sure window frame is structurally sound enough to support the unit and hold the fasteners.

ELECTRICAL DATA

Voltage Supply – For satisfactory results, when unit is running, the voltage must always remain within ± 10% of the rated nameplate voltage. Unit operation above or below the ± 10% of nameplate voltage can damage unit. Check voltage supply at outlet and resolve any wiring problems before proceeding with the installation.

Grounding – Unit must be grounded. It is grounded thru the service cord plug and matching receptacle when branch circuit is grounded. Be sure branch circuit is grounded.

Consult local electrical codes for recommended grounding methods.

Unit is recommended for use on a grounded power supply only.

Receptacles – The field-supplied outlet receptacle must match the plug on the service cord of your room air conditioner and must be located within 6 ft of the unit. The recommended 15-amp, 125-volt receptacle shown is an Arrow-Hart 5251, P & S 5251, Hubbell 5251 or GE4064-1. Wiring must comply with all local and national electrical codes.

Service Cord and Branch Circuit Wire Size

Individual branch circuit wire size is 14 AWG. *Do not splice cords or attempt to operate room air conditioner on undersized extension cord.*

Fuse Size – Use a 15-ampere dual element (time delay) "Fusetron" or equivalent, or 15-ampere circuit breaker.

STEP-BY-STEP INSTALLATION

This model is for installation in double-hung windows 25 1/4-in. to 40-in. wide and a minimum sash opening of 14 1/2 in. as shown in Fig. 1.

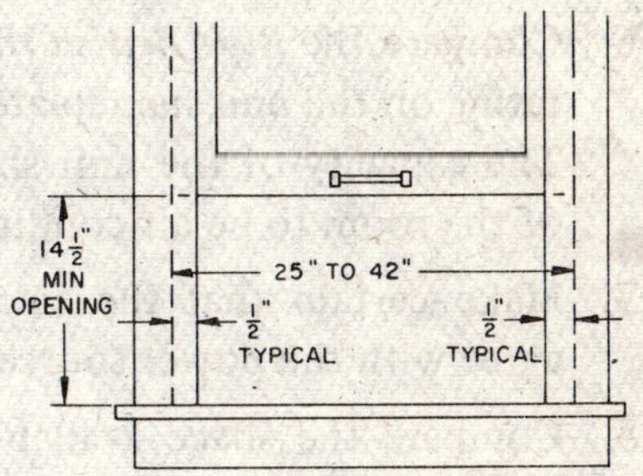

Fig. 1 – Window Dimensions (Double-Hung)

Step 1 – Install unit support bracket on window sill.

NOTE: The use of this bracket is required for supporting unit and to prevent unit from falling out of window when the sash is raised.

a. Screw leveling screw into cage nut on bottom of bracket.
b. Position bracket on sill with side of bracket 3/4 in. to left of window center line; open end against window stool (inside sill) as shown in Fig. 2.
c. Fasten bracket loosely to sill with screws provided.
d. Adjust leveling screw until notch in back edge of bracket is level with stool as in Fig. 2. Fasten bracket securely to sill.

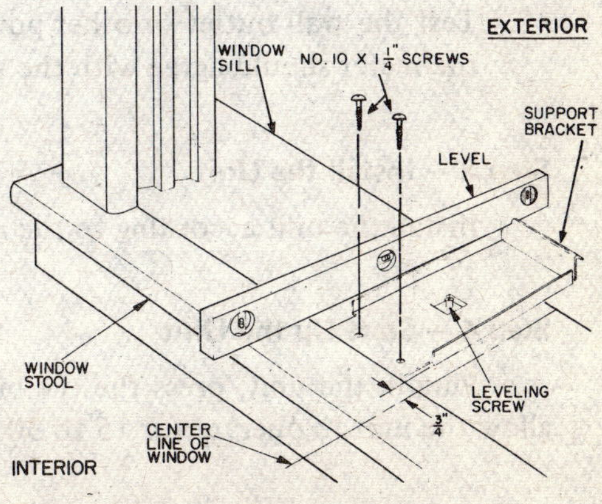

Fig. 2 – Installing Support Bracket

Fig. 7-4 Typical installation instructions

the instructions. If there is a packing list, carefully compare the items shipped against the items on the list, and report any missing parts so that a claim can be filed.

5. Read the manufacturer's instructions carefully and be sure you understand them.
6. Compare the *Btu* (*British thermal unit*, the standard unit for measuring heat) rating on the unit nameplate with the required Btu cooling load for the room. The capacity of the unit should be large enough to handle the cooling load of the room to be air conditioned.
7. Make certain that the electrical voltage and phase on the unit nameplate agree with the power source available.
8. Compare the space available for unit installation with the unit dimensions.
9. Determine the location of the unit. Most units operate best when installed on the side of the house that has the most shade at the hottest time of day.
10. If the unit is to be installed in a window, use a suitable windowsill support bracket kit (shipped with the unit). If the unit is to be installed through the wall, complete the proper wall framing as specified in the manufacturer's instructions.

Step 2 – Test the Power Source

1. Before plugging the unit into the wall outlet, make sure that the power cord is properly connected at both ends and in good working order by testing for continuity with an ohmmeter. An unbroken, continuous flow of electricity should result. If the ohmmeter shows a broken flow, determine the cause and make proper repairs by tightening the connection or replacing the cord.
2. Test the wall outlet or other power source with a voltmeter. The reading on the meter should agree with the voltage listed on the unit nameplate.

Step 3 – Install the Unit

Install the unit according to the manufacturer's instructions.

Step 4 – Start Up the Unit

Plug in the unit, press the ON button, figure 7-5, and allow the unit to operate for 15 to 30 minutes.

Fig. 7-5
ON-OFF button

Step 5 – Measure the Discharge Air Temperature

With the unit operating, use a pocket thermometer to measure the temperature of the cold air flowing from the discharge grille. The discharge temperature should be in the range of 40 F to 50 F for proper cooling.

Step 6 – Measure Return Air Temperature

Measure the temperature of the air flowing back or returning to the unit at the return air grille. The thermometer should read close to room temperature since the return air (from the room to the unit) reflects air temperature in the room.

Step 7 – Clean Up the Job Site

Clean up the space around the installed unit. Vacuum or wipe up dust on the floor under the unit. Pick up scrap material or other debris that may have fallen outside on the ground under the unit, and dispose of it properly.

Step 8 – Give Instructions to the Owner

Give the owner the instruction booklet and discuss the following points:

1. How to Operate the Unit
   a. Start the unit before comfort conditions are needed in the room. When there is heat in a room, and hot outside temperatures, a longer cooling-down period is usually required. Judgment and experience will help determine when to start the unit.
   b. Follow the manufacturer's instructions for operating the control buttons (fans, high-cool/low-cool exhaust).
   c. Close drapes, pull down shades, and otherwise prevent heat from the sun from coming through the windows in the room.
   d. Close the doors and windows in the room. If they are left open, cool air escapes and the room will not be properly cooled.
2. How to Care for the Unit Properly
   a. Most window units have painted surfaces that should be cleaned as frequently as required with a mild detergent and water.
   b. During the start-up period, humid conditions in the room may cause tiny water droplets to form on the grille. Continue to operate the unit to lower the humidity to the proper level.
   c. Clean the filter as frequently as necessary to maintain the proper airflow. Most filters are located behind the front grille and should be removed, cleaned, and replaced according to the manufacturer's instructions, figure 7-6.

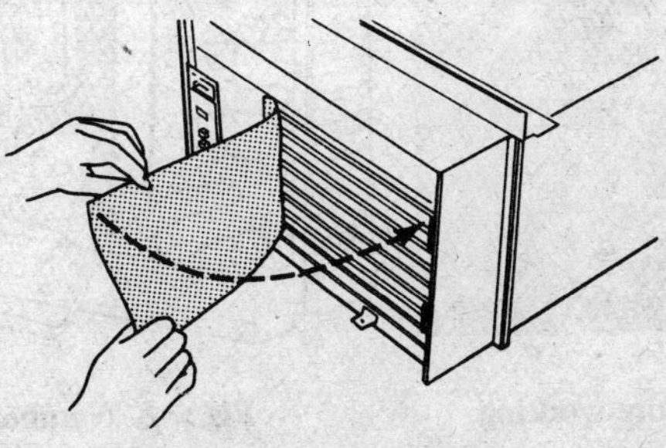

Fig. 7-6 Filter in place

d. Solvents in insect sprays are harmful to the metal and other surfaces of the unit. Do not use these near the unit, especially when it is operating

3. Before Requesting Service
   a. Be sure the unit power cord is plugged into the wall outlet.
   b. If proper cooling is not being maintained:
      (1) Clean the filter.
      (2) Check the fuse and replace or reset it if it has blown.
      (3) Make certain the thermostat is set correctly.
      (4) Determine that a general power failure has not occurred.
   c. After performing any or all of the above, wait a few minutes and test-run the unit.

GENERAL SERVICING RULES AND TECHNIQUES

1. Always shut off the power to the unit before servicing it, figure 7-7.
2. Keep a fire extinguisher handy when using a torch.
3. Follow all local and national safety codes.
4. When disconnecting electrical wires from terminals, be sure to identify each wire and terminal by color or number, figure 7-8, so they can be reconnected quickly and accurately.
5. Follow the manufacturer's instructions when charging out compressors or capacitors. This operation is usually done by an experienced master mechanic, not by a beginner.
   a. When hermetic compressor motors burn out, the refrigerant system becomes contaminated and must be properly flushed and purged. Usually

Fig. 7-7 Turn off all power before working on electrical equipment.

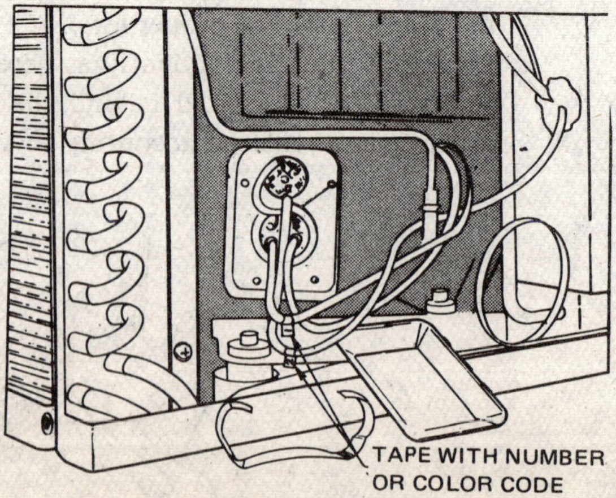

Fig. 7-8 Number or color code wires to terminals when disconnecting

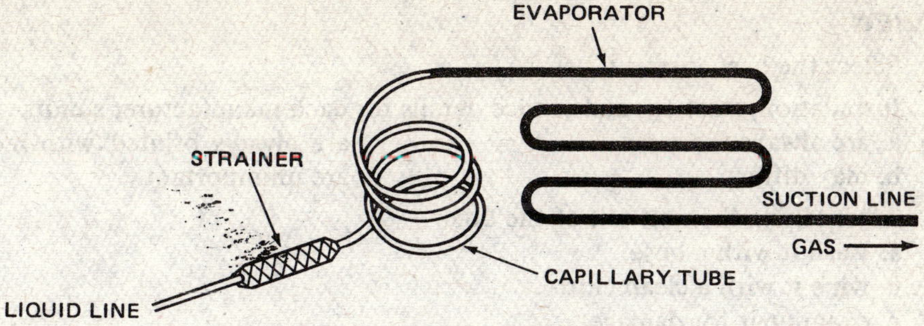

Fig. 7-9 Capillary tube and strainer

a new system strainer and capillary tube, figure 7-9, are installed at the same time.

b. Cut the refrigerant lines to and from the compressor at convenient points so that soldered reconnections will be easy to make.

6. Perform maintenance annually or more frequently if necessary.
   a. Repaint rust spots, tighten loose electrical and piping connections, repitch the unit for proper drainage, and clean drain holes at least once a year.
   b. Clean the coils. They can be cleaned with a vacuum cleaner, a suitable brush, or compressed air. Dry coils clean more easily than wet coils.
   c. Check all controls, including the ventilation door, for proper operation.

GENERAL SERVICE PROBLEMS

On Every "No-Cooling" Service Call:

Check For:	Remedy:
1. Power to unit with voltmeter.	1. If power not available, determine the cause and correct the problem.
2. Loose electrical and piping connections.	2. Tighten all connections as necessary.
3. Blown fuses.	3. If blown, reset or replace.
4. Dirty filters.	4. Clean and replace as necessary.
5. Serious restriction of airflow through indoor and outdoor coil.	5. Remove restrictions and clean coils as necessary.
6. Fan performance. (If air is below comfort level.)	6. Look for loose connections, restrictions, or weak power supply. Correct each as necessary.
7. Defective ON - OFF and other control switches.	7. If defective or seriously worn, replace.
8. Iced-over indoor coil.	8. Add refrigerant. Ice on coil usually indicates low refrigerant charge.
9. Power to compressor.	9. If no power, test for defective capacitor, overload switch, relay or frozen (seized) compressor. Correct or replace as necessary.

REVIEW

Select the best answer.

1. Installation, start-up, and service details for each manufacturer's units
   a. are always the same.
   b. may differ.
   c. are always printed with red ink.
   d. are unimportant.
2. After taking the unit out of the box
   a. wash it with a hose.
   b. wipe it with a clean cloth.
   c. examine it for damage.
   d. keep it outside until ready to install.
3. Compare the Btu rating on the unit nameplate with
   a. ARI ratings.
   b. ATMS ratings.
   c. compounded ratings.
   d. the required cooling load for the room.
4. Compare the unit nameplate electrical voltage with
   a. ARI ratings.
   b. ATMS voltages.
   c. the available power-source voltage.
   d. voltage required on a previous, similar installation.
5. Voltmeter reading at the wall outlet or power source should agree with
   a. ARI.
   b. NEMA.
   c. the voltage listed on the unit nameplate.
   d. the voltage of a previous similar installation.
6. Measure discharge air temperature
   a. with a pocket thermometer.
   b. when the unit is not operating.
   c. with a hygrometer.
   d. by slowly moving your hand across the airflow.
7. Instruct the homeowner to call for service
   a. as soon as the unit stops operating.
   b. only after checking the unit power cord, cleaning the filter, checking the fuse, etc.
   c. late in the evening.
   d. after an electrical storm.
8. Before servicing the unit
   a. turn on the lights.
   b. call for assistance.
   c. wipe dirt off the unit.
   d. turn off the power.
9. Charging out a burned-out compressor should be done
   a. by an experienced mechanic only.
   b. by an apprentice only.
   c. by the homeowner.
   d. only at night.
10. On every service call, check for
    a. weak fins.
    b. weak tubes.
    c. the number of trees near the unit
    d. the correct power to the unit.

8
Installing Central Residential Cooling Coils

OBJECTIVE

After studying this unit, the student will be able to

- install and service a residential cooling coil in the attic.

INTRODUCTION

A cooling coil is usually enclosed in a casing, as illustrated in figure 8-1. It is used as a module with other components such as fans, condensing units, and furnaces to make up a system. Several residential or small commercial applications are possible. The coil can be installed in the horizontal duct run downstream from a horizontally installed attic furnace, or in the plenum of a vertically installed upflow furnace. The coil can also be installed in the plenum of a vertically installed downflow furnace, or in a straight run of horizontal attic duct ahead of (upstream from) a fan section module. This unit outlines the last of these applications. Keep in mind that a condensing unit is required in addition to the cooling coil to complete the system. In this unit, only cooling-coil installation is discussed.

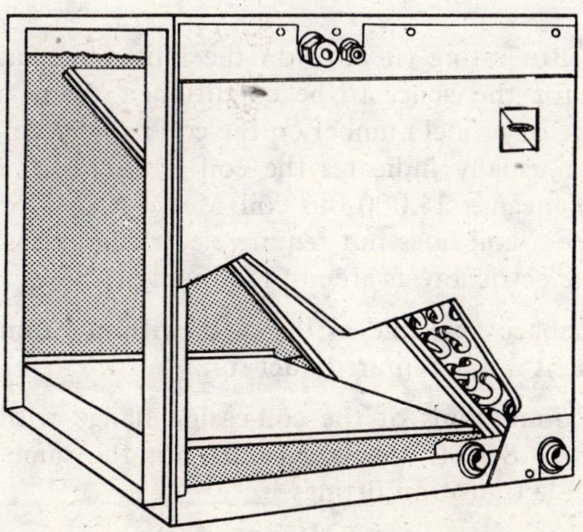

Fig. 8-1 Cooling coil in casing

Although installation, start-up, and service details may differ with each manufacturer, the general procedures presented here apply to most cooling coil brands. Specific details are usually covered in the manufacturer's instructions available with each unit. Cooling only is discussed here. Unit 15, "Duct Sizing" and Unit 16, "Residential Air Conditioning" in the text *Principles of Air Conditioning* may be useful references if a proper cooling effect is not achieved. Also refer to the chart "General Service Problems" in this unit.

PROCEDURE

Step 1 – Getting Ready to Install the Coil

1. Take the coil and casing out of the box, crate, or carton in which it was shipped.
2. Examine the coil and casing for damage such as bent panels, paint scratches, crushed or bruised coil fins, and damaged piping connections. File a claim if necessary. Such claims are usually made to the shipping company.
3. Find the manufacturer's installation instructions, figure 8-2, in one of several possible locations, such as in an envelope or plastic bag taped or tied to the coil, in the shipping carton, or inserted inside the coil casing.
4. Depending upon the manufacturer's policy and shipping practices, a list of items packed in the shipping carton may be itemized in the front section of the instructions. If there is a list, carefully compare the items shipped against the items on the list. Since a cooling coil is quite simple, the only items shipped may be the coil with casing, and a bag of nuts and bolts or sheet metal screws.
5. Read the manufacturer's instructions carefully and make sure you understand them.
6. Compare the Btu rating (if any) on the unit nameplate with the required cooling load for the space to be conditioned. Some manufacturers show only a product or model number on the coil nameplate. The last part of the model number usually indicates the coil size in Btu. For example, Model AAA024 may mean a 24,000-Btu coil; Model AAA036 may mean a 36,000-Btu coil. Since a coil does not require electric hookups, verification of voltage and other electrical tests are not required.
7. Compare the space available in the attic with the dimensions of the coil. The unit must fit in the required duct run.
8. Compare the dimensions of the coil-casing flange with the duct-flange dimensions, figure 8-3. If the flanges are not the same size, plan ahead to fabricate suitable transition fittings.
9. Determine the location of the unit. It should be positioned in the duct run ahead of the system fan. The coil and casing should rest on a support on the attic joists. A wooden platform is usually strong enough to bear the weight of the unit. If the duct run and the system fan section are hung from the

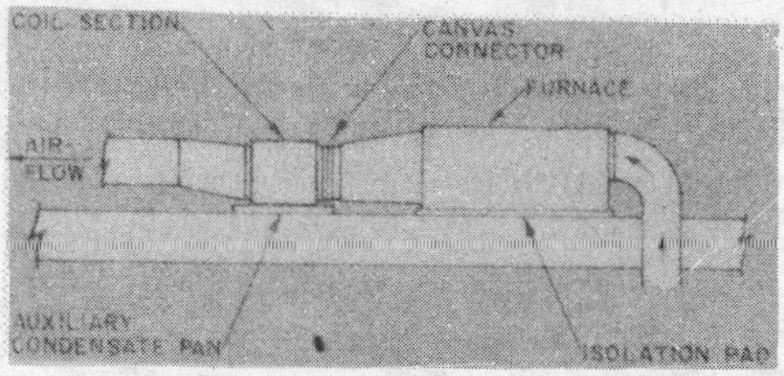

Fig. 5 — Typical Attic or Overhead Crawl Space Coil Installation

REFRIGERANT PIPING

Thermal expansion valve, distributor and nozzle are factory installed. Field fabricate liquid and suction piping thru front access panel or knockouts on end panels of casing. Sweat connections to field piping. See Fig. 1 for connection sizes and location.

Thermal expansion valve and liquid line must be field installed. Figure 6 shows typical field-fabricated piping to the thermal expansion valve. Equalizer line to the thermal expansion valve is factory piped but must be field installed.

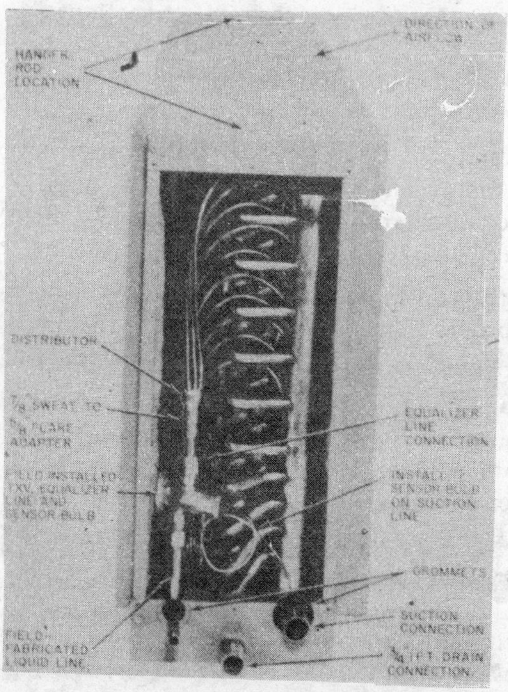

Fig. 6 — Typical Field-Fabricated Piping to Thermal Expansion Valve

Refrigerant lines are designed for sweat-type connections on suction line and flared-type connections on liquid line to thermal expansion valve. See Fig. 1 for connection sizes and location.

CAUTION: Mount expansion valve sensor bulb with capillary tube extending from top of bulb. Locate sensor bulb outside of unit casing and away from liquid traps or pockets in the suction line to avoid throttling in the thermal expansion valve. Ensure that sensor bulb is properly installed before operating system to avoid slugging of liquid to the compressor.

SUCTION LINE PIPING

1. Remove screws holding panels to front of unit and remove panels.
2. Set small access panel in place to ensure proper piping.
3. Pull thermal expansion valve sensor bulb and capillary thru suction line opening of panel.
4. Extend suction line approximately 12 in. from casing and connect as shown in Fig. 7 for easy removal of access panel.
5. Approximate location of the sensing bulb with respect to suction line as shown in Fig. 7 and clean suction line at the contact area.
6. Attach thermal expansion valve sensor bulb to suction linc with thc straps provided.
7. Insulate suction line and thermal expansion valve sensor bulb. Ensure that insualtion extends to unit casing.

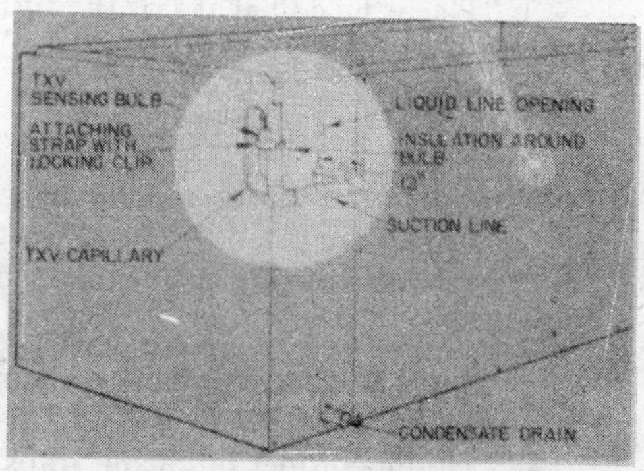

Fig. 7 — Recommended Suction Line and Sensing Bulb Installation

LIQUID LINE PIPING

1. Measure and install piping and elbows required for joining thermal expansion valve inlet port thru access panel to the external liquid line.
2. Replace access panel and secure with screws.

Fig. 8-2 Typical installation instructions

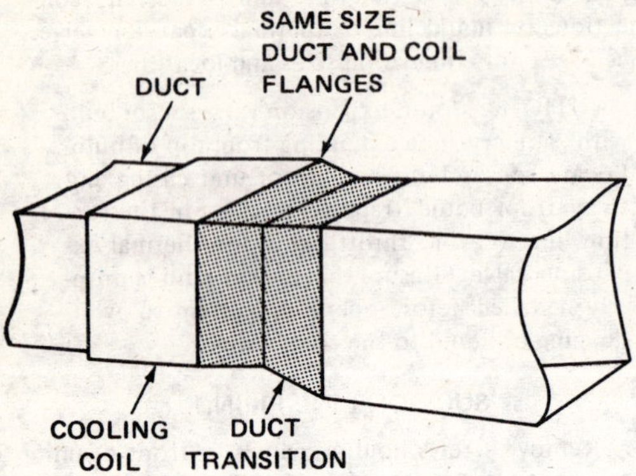

Fig. 8-3 Mating coil and duct flanges

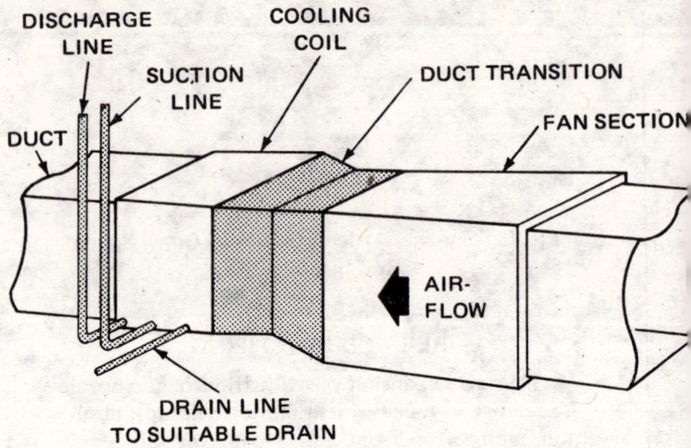

Fig. 8-4 Typical horizontal duct installation

attic rafters, the coil and casing should also be suspended from the roof with suitable hanging straps.

10. Be sure to follow the manufacturer's instructions carefully when determining the location of condensate drain lines and drain pans. Some horizontal units must be positioned properly to insure that the correct drain connections are used.

Step 2 – Install the Coil and Casing

1. Place the coil ahead of the system fan in the attic duct run. Follow the manufacturer's instructions to be sure that drain connections are properly positioned, figure 8-4. Some units have two sets of primary and secondary drains; it is important to position the correct set of drains at the bottom of the unit.
2. Connect the attic-duct flanges to the coil-casing flanges. If the two flanges are not the same size, fabricate a proper transition fitting. Be sure the drain pan and coil are not pierced or damaged during sheet-metal fabrication.
3. Install drainpipes so that water condensing from the coils will drain to the run-off location without damaging property. Running the condensate lines to an open drain or sump is recommended. Use good piping practices, including proper pitch and traps, when installing drain lines. Follow the manufacturer's instructions.
4. Connect the primary and secondary drain lines. If the manufacturer's drain line and drain pan instructions are not followed, serious water damage to the living area below the coil can result. Most manufacturers recommend a secondary drain line for added protection when the unit is installed above living areas.
5. Connect the condensing-unit refrigerant suction and discharge lines to the coil. Connections at the coil may be either sweat or quick-connect fittings. In either case, follow the manufacturer's instructions to insure clean lines and clean refrigerant.

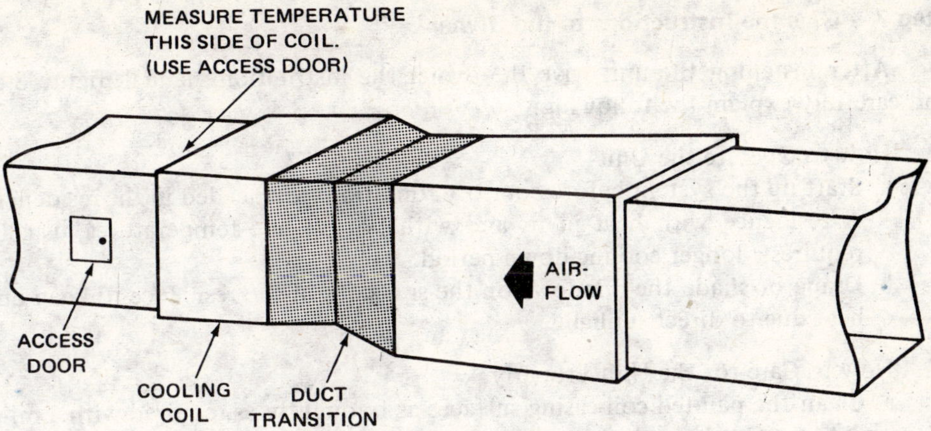

Fig. 8-5 Measuring downstream coil temperature

Step 3 – Start Up the Air-conditioning System

Although only cooling-coil installation is discussed in this unit, the entire system must be started up to test the coil operation. Follow the manufacturer's system start-up instructions (usually supplied with the condensing unit).

Step 4 – Measure the Discharge Air Temperature

With the system operating, measure the cold air flowing downstream from the coil, figure 8-5. If thermometer access to the duct run is not possible, measure the air temperature at a discharge grille in the conditioned space. The temperature in the discharge duct run near the coil should be in the 40 F to 50 F range for proper cooling results. Higher temperatures will be measured at the discharge grille in the conditioned space.

A second method of determining performance is to measure air temperature just ahead of (upstream from) the cooling coil and just after (downstream from) the coil. Temperature differences between the two measurements should range from 15 F to 25 F, depending upon the specific coil. If more cooling is required, increase the fan speed. Decrease the fan speed for less cooling.

Step 5 – Measure the Return Air Temperature

Use a pencil (pocket) thermometer to measure the air temperature flowing back to the unit at the return air grille. The thermometer should read close to room temperature since return air from the room should reflect air temperature in the room.

Step 6 – Clean Up the Job Site

Clean up the space around the installed coil and around the drain line and refrigerant line connections.

Step 7 – Give the Instructions to the Owner

After installing the unit, give the owner the instruction manual, figure 8-6, and carefully explain the following:

1. How to Operate the Unit
   a. Start up the system before comfort conditions are needed in the residence. A residence with heat in it, and with hot outside temperatures, usually requires a longer cooling-down period.
   b. Drape or shade the windows on the sun sides of the residence to keep out heat due to direct sunlight.
2. How to Care for the Unit Properly
   a. Clean the painted coil-casing surfaces as frequently as required with a mild detergent and water.
   b. Inspect and clean the drain pan and drain outlets periodically to insure the free flow of condensate through both the primary and secondary drain lines.
   c. Inspect and clean the coil surface periodically to keep the coil-fin surfaces clean. Coils can be cleaned with a vacuum cleaner, a suitable brush, or compressed air. Dry coils clean better than wet coils.
   d. Solvents in insect sprays are harmful to the metal surfaces of the coil. Do not use or store them around the coil, especially when the system is operating.

GENERAL SERVICE PROBLEMS

On Every "No-Cooling" Service Cal'

Check For:	Remedy:
1. Serious restriction of airflow through indoor cooling coil.	1. Remove restrictions and clean coils as described under "How to Care for the Unit Properly," Step 7.
2. Iced-over cooling coil.	2. Add refrigerant to system. Ice on coil usually indicates low refrigerant charge.
3. Refrigerant loss at coil.	3. Repair leaks as required and recharge the system.
4. Cooling load estimated incorrectly, or improper duct or coil size.	4. Refer to Units 15 and 16 in the text *Principles of Air Conditioning* for proper cooling-load estimating procedures. Then check the coil manufacturer's selection book for proper coil sizes. In *Principles of Air Conditioning,* Units 11 and 12 (regarding heating- and cooling-load estimating) may also be helpful.

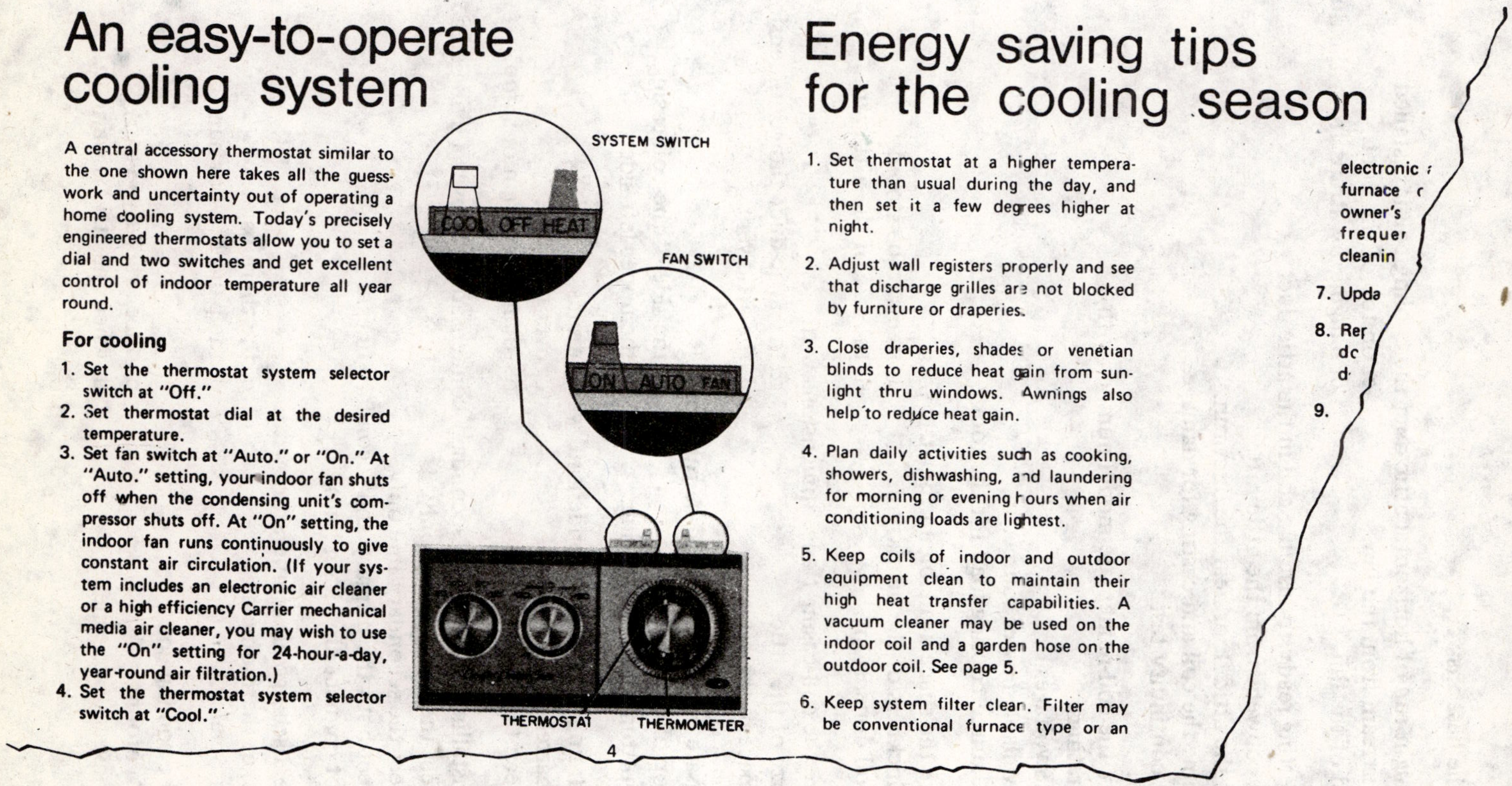

An easy-to-operate cooling system

A central accessory thermostat similar to the one shown here takes all the guess-work and uncertainty out of operating a home cooling system. Today's precisely engineered thermostats allow you to set a dial and two switches and get excellent control of indoor temperature all year round.

For cooling

1. Set the thermostat system selector switch at "Off."
2. Set thermostat dial at the desired temperature.
3. Set fan switch at "Auto." or "On." At "Auto." setting, your indoor fan shuts off when the condensing unit's compressor shuts off. At "On" setting, the indoor fan runs continuously to give constant air circulation. (If your system includes an electronic air cleaner or a high efficiency Carrier mechanical media air cleaner, you may wish to use the "On" setting for 24-hour-a-day, year-round air filtration.)
4. Set the thermostat system selector switch at "Cool."

4

Energy saving tips for the cooling season

1. Set thermostat at a higher temperature than usual during the day, and then set it a few degrees higher at night.
2. Adjust wall registers properly and see that discharge grilles are not blocked by furniture or draperies.
3. Close draperies, shades or venetian blinds to reduce heat gain from sunlight thru windows. Awnings also help to reduce heat gain.
4. Plan daily activities such as cooking, showers, dishwashing, and laundering for morning or evening hours when air conditioning loads are lightest.
5. Keep coils of indoor and outdoor equipment clean to maintain their high heat transfer capabilities. A vacuum cleaner may be used on the indoor coil and a garden hose on the outdoor coil. See page 5.
6. Keep system filter clean. Filter may be conventional furnace type or an

electronic
furnace
owner's
frequer
cleanin

7. Upda
8. Rer
dc
d
9.

Fig. 8-6 Typical owner's instructions

REVIEW

Select the best answer.

1. In a straight run of horizontal attic duct the cooling coil is installed
   a. downstream from the fan.
   b. ahead of the fan.
   c. on top of the duct.
   d. hanging by straps from the duct.

2. When getting ready to install a coil in the attic duct
   a. clean cobwebs from the attic rafters.
   b. paint the attic floor near the duct run.
   c. examine the coil and casting for damage.
   d. run a continuity test.

3. Compare the coil-nameplate model number with
   a. ARI ratings.
   b. ATMS voltages.
   c. the available power source.
   d. the required cooling load for the room.

4. Compare the space available in the attic with
   a. the thickness of the insulation.
   b. the coil dimensions.
   c. items to be stored.
   d. the size of the attic vent.

5. Compare the coil-casing flange dimensions with
   a. the size of the attic.
   b. the space in the attic.
   c. the attic-duct flange dimensions.
   d. the attic vent.

6. Condensate drain lines should be run from
   a. the top of the coil casing.
   b. the bottom of the coil casing.
   c. the side of the coil casing.
   d. the upper left side.

7. Install drain pipes so that condensate water drains
   a. to a barrel.
   b. between the walls.
   c. over the edge.
   d. to an open drain or sump.

8. When installing primary and secondary condensate lines over a living area
   a. connect the primary line only.
   b. connect the secondary line only.
   c. connect both the primary and secondary lines.
   d. do not connect any drain lines.

9. Measure discharge air temperature with
   a. a pocket thermometer.
   b. a hygrometer.
   c. the thermostat.
   d. the palm of your hand.

10. Clean the cooling coil
    a. with a broom.
    b. with a vacuum cleaner.
    c. once every five years.
    d. when the attic is warmest.

9
Installing Central Residential Cooling Systems

OBJECTIVE

After studying this unit, the student will be able to

- install a central residential air conditioning, split system.

INTRODUCTION

In a *split system*, the condensing unit is installed outdoors and the cooling coil is installed indoors. In the application used in this unit, the cooling coil is to be located in the existing furnace plenum, figure 9-1.

Although installation, start-up, and service details may differ from manufacturer to manufacturer, the general procedures presented here apply to most brands. Specific details are usually covered in the manufacturer's instructions available with each unit.

PROCEDURE

Step 1 – Getting Ready to Install the Outdoor Condensing Unit

1. Move the unit to the installation site and slide it out of the box, crate, or carton in which it is packaged.
2. Examine the unit for shipping damage such as bent panels, scratches, loose or bent fans or fan housing, crushed or bruised coil grilles, and cracked casings, and file a claim if necessary.
3. Find the manufacturer's instructions taped or tied to the unit, in an envelope, in the shipping carton, or inserted in a convenient slot in the unit. Usually, the instructions are taped or tied to the top of the unit.
4. Some manufacturers include a list of items packed in the shipping carton in the front section of the instructions. If there is a list, carefully compare the items actually shipped with the items on the list. Report any missing parts and file a claim if necessary.

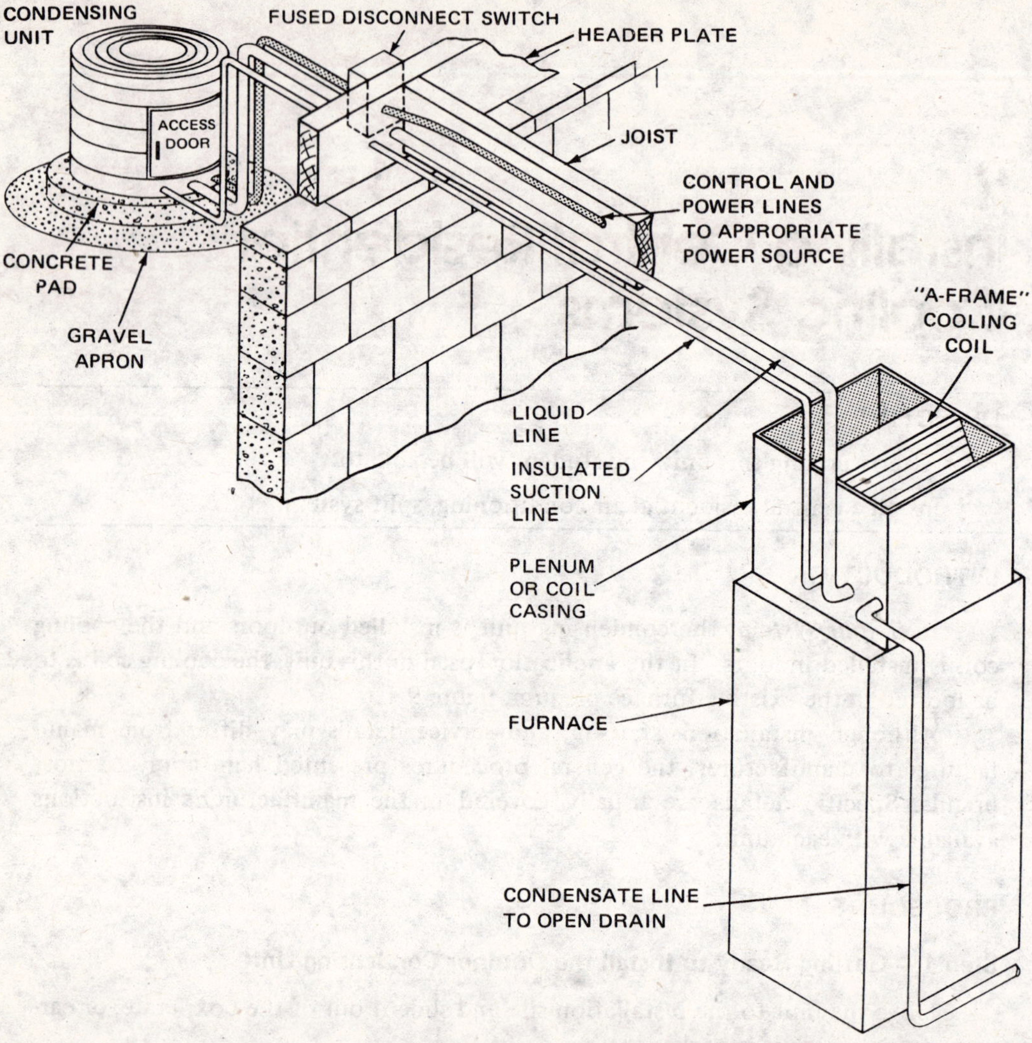

Fig. 9-1 Typical split system

5. Study the manufacturer's instructions carefully and be sure you understand them.
6. Compare the Btu rating listed on the unit nameplate with the required cooling load for the space to be conditioned. The Btu capacity of the unit should be large enough to handle the cooling load.
7. Make certain that the electrical voltage and phase shown on the nameplate agree with the available power source.
8. Compare the space available for unit installation with the unit size and the space required for servicing.
9. Determine the location of the unit. Most units operate best when installed on the side of the house that has the most shade at the hottest time of day.

The location of the outdoor unit must also take into account the unit's sound level. If the shadiest spot is on the property line and near neighbors' living areas, select a compromise location, or erect an effective sound barrier such as a wall to comply with local noise ordinances.

Step 2 – Mount the Outdoor Base Pad and Condensing Unit

Snow is not a factor in this application.

1. Install a concrete mounting pad on a gravel base located where rain or ice from the roof will not damage the unit. Follow the manufacturer's instructions for the procedure.
   a. Use a round pad for a round condensing unit, figure 9-2, and a square pad for a square unit. A pad compatible with the style of the unit gives the best appearance.
   b. The construction of the pad should be in proportion to the weight of the unit. Be sure the weight of the pad agrees approximately with the manufacturer's recommendations.
   c. Install tie-down bolts as indicated in the unit instructions; two bolts are usually sufficient.
   d. Refer to the following (item 2) for spacing considerations.
2. Place the condensing unit on the cement base pad. Be sure to allow enough space for making refrigerant piping runs and connections and for running the power and control wiring to the terminals inside the unit. Enough clearance for adequate airflow must also be allowed. If the service door or panel is positioned near a wall, allow sufficient space for easy access. Follow the manufacturer's instructions in all of these spacing considerations.
3. With the unit in place on the base pad, determine a suitable point for drilling holes in the header plate or foundation wall for the refrigerant piping and for the control and power wiring, figure 9-3.

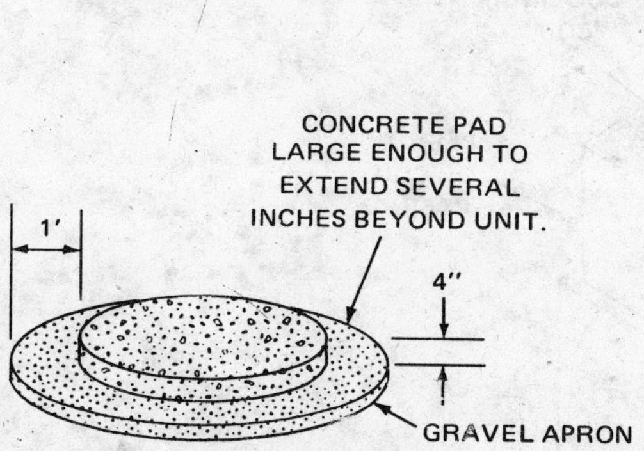

Fig. 9-2 Circular concrete pad

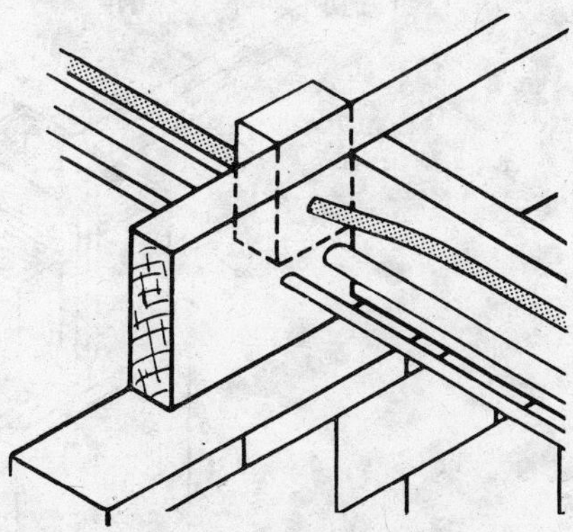

Fig. 9-3 Holes required for piping and wiring

4. Slide the unit away from the wall of the base pad if necessary, and drill holes as required. The holes should be just large enough to accommodate the piping and wiring so that minimal caulking will be required later.

5. Reposition the unit and anchor it to the mounting bolts. If space is limited, it may be practical to install the wiring and the fused disconnect switch before repositioning the unit.

Step 3 – Mount the Cooling Coil in the Furnace Plenum

If the furnace and cooling coil are designed by the same manufacturer, the coil should fit the furnace top with a minimum number of flanges, or none, required. If the coil and furnace are of different brands, a flange arrangement may be required as a base to hold the coil in position on the furnace, figure 9-4. In either case, follow the manufacturer's instructions for installing a flange that accommodates the dimensions of the coil.

1. Move the coil to the furnace area and remove it from the box, crate, or carton in which it was shipped.
2. Examine the coil for damage such as scratches, crushed fins, and loose mountings. File a claim if necessary.
3. Locate the manufacturer's instructions and study them thoroughly.
4. Remove the necessary furnace plenum ductwork and measure the opening at the base of the plenum. Compare the size of the plenum base opening with the size of the coil base to make sure the coil fits securely on top of the furnace plenum base.
5. If the coil base is smaller or larger than the plenum base opening, install supporting flanges on the plenum base according to the manufacturer's instructions.

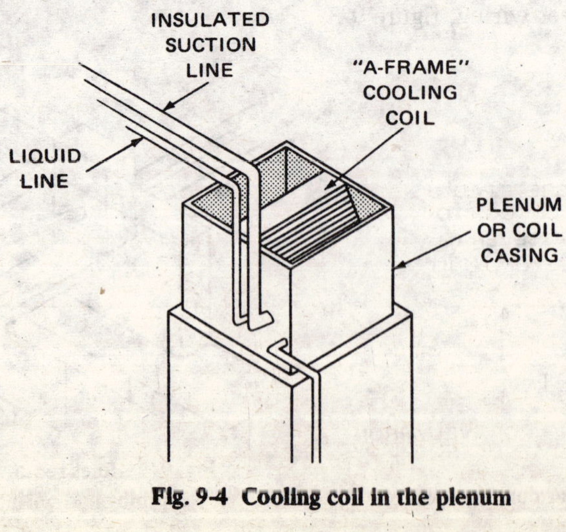

Fig. 9-4 Cooling coil in the plenum

6. Mount the cooling coil by placing it on the base of the furnace plenum. Be sure the refrigerant and coil drain-line connections are positioned properly for easy hook-up.

   Note: If the coil is shipped with a coil casing, skip item 7.

7. Place the plenum casing over the coil temporarily and position it on the furnace. Determine where the refrigerant and drain-line connections from the coil will enter the plenum, and mark the points on the plenum casing. Remove the casing and cut or punch appropriate size holes for the refrigerant and drain lines. Replace the casing. A removable panel with appropriate holes may also be used.

Step 4 – Install the Refrigerant Piping

Refrigerant piping connections can be made by using quick-connect, speed-connect, compatible, flared, or sweat fittings. Some manufacturers supply piping or tubing packages of varying lengths with speed, or compatible fittings; these are available precharged with refrigerant. Piping used with flared or sweat fittings is not precharged. If flared or sweat fittings are used, additional refrigerant should be added to the system according to the manufacturer's instructions.

Perform the following procedures when installing refrigerant piping:

1. Read the manufacturer's instructions carefully and determine the proper size and length of tubing for the equipment being installed. For this application, select a precharged tubing package as close to the required length as possible.

2. Do not cut precharged tubing. Instead, carefully curl it into loops to take up any excess tubing. Even when tubing is not precharged, carefully examine the ends of the tubes before cutting them to the proper length. They may have been specially prepared to fit the connection and will no longer fit if the end is removed.

3. Carefully unroll the tubing package and run it from the condensing unit connection through the holes in the header plate to the cooling coil connection on top of the furnace. Keep in mind that with continuous tubing, bends should be made only by following acceptable tube-bending practices. Do not crimp or kink the tube by making a bend that is too sharp.

4. Install isolation hangers to prevent vibration noises from being transmitted to joists or beams.

5. Make quick-connect and compatible fitting connections at the condensing unit and cooling coil according to the manufacturer's instructions.

6. Install the drain line to an open drain.

Step 5 – Install the Wiring

All wiring runs, connections, and wire size selections should be performed or supervised by an experienced, properly qualified electrician.

In this unit, the wiring instructions are based on the assumption that the furnace fan and motor are adequate to handle the heating or cooling requirements; the furnace fan and motor have been presized for cooling duty.

A two-stage heating and cooling thermostat must be substituted for the heating-only thermostat installed in the existing system.

Line Voltage Wiring to the Outdoor Unit. When installing line voltage wiring to the compressor in the outdoor unit, complete the following steps:

1. Study the manufacturer's instructions thoroughly and be certain that all wiring complies with local and national electrical codes and ordinances.
2. Install a properly sized fused disconnect switch on the wall near the outdoor unit, figure 9-5.
3. Run line voltage wiring with the appropriate volt-ampere capacity (as specified in the manufacturer's instructions) from the power source to the outdoor fused disconnect switch.
4. Connect the wiring at the power source and at the disconnect switch, using proper safety precautions, figure 9-6. Be sure that the ground wire is connected at both ends and that the disconnect switches are in the OFF position at both the power source and the outdoor disconnect.
5. Run the power wiring, including the ground wire, from the outdoor disconnect switch to the outdoor unit, and connect it at the proper terminals as indicated in the manufacturer's instructions.

Low-voltage Control Wiring. To install low-voltage control wiring from the thermostat to the furnace and outdoor unit, use the following general procedures:

1. Study the manufacturer's control wiring instructions thoroughly and be certain you understand them.
2. Shut off power to the furnace.
3. Remove the existing heating-only thermostat and subbase.
4. Following the manufacturer's wiring instructions closely, run the low-voltage control wiring from the furnace or control-box transformer to the proper outdoor unit contacts, figure 9-7.
5. Run the low-voltage control wiring from the thermostat to the outdoor unit.
6. Install a heating-cooling thermostat.
7. Connect all control wiring to the terminals as indicated in the manufacturer's instructions.

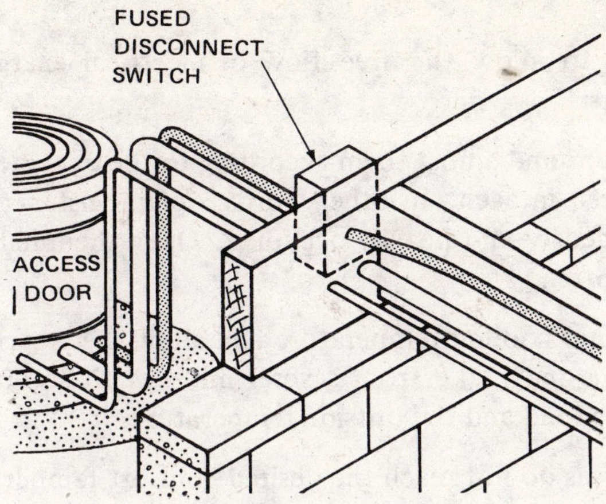

Fig. 9-5 Fused disconnect switch

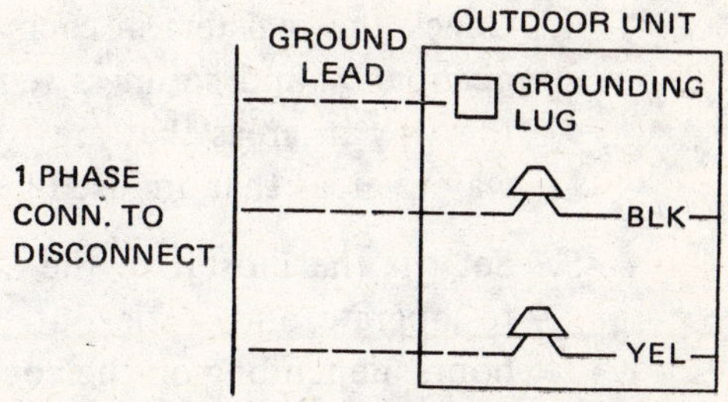

Fig. 9-6 Typical power connections to outdoor unit

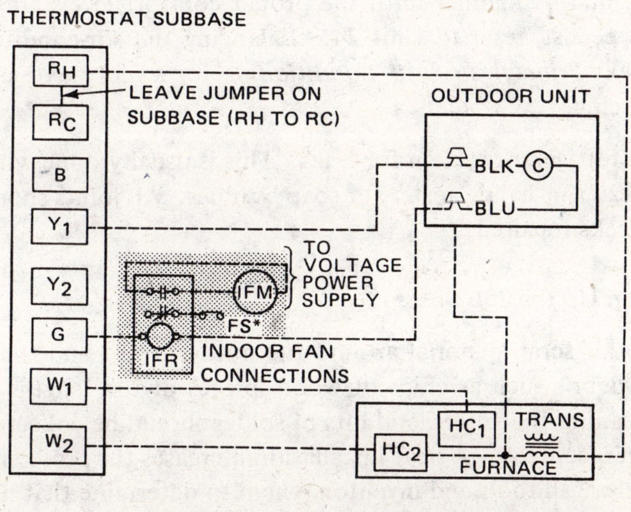

Fig. 9-7 Typical control wiring connecting furnace, outdoor unit and thermostat subbase

Step 6 – Check Out All Piping and Wiring

Check all piping and wiring circuits and connections to make certain they agree with the manufacturer's instructions and with local and national electrical codes and ordinances.

Step 7 – Start Up the System

1. Check the furnace filter and replace or clean it if it is dirty.
2. Open the suction and discharge valves according to the instructions to allow the free flow of refrigerant.

3. Close all disconnect switches to allow the free flow of electrical energy through the system components.

4. Check the registers in each room and adjust them according to the expected occupancy and coolness required in each. In other words, open registers all the way in areas that are frequently and heavily occupied. Open them halfway in areas that are used less often.

5. Set the thermostat to the desired cooling temperature and allow the system to operate and settle out. This may take from several minutes to several hours, depending on the cooling load and the outside temperature.

6. If some of the less occupied areas do not reach the desired comfort temperature, readjust the registers to allow more cool air to enter the warmer areas. For example, move the wide-open registers to three-quarters open. This should cause more air to flow into the warmer areas. Continue to adjust the registers in this manner until the proper comfort levels are reached. If difficulties persist, refer to Unit 24, "Balancing the Air-conditioning System," in the text *Principles of Air Conditioning* for a more extensive approach to air balancing.

7. Test the refrigerant system for leaks. This is usually done with either an electronic tester, a halide tester, or soap bubbles. All joints should be examined and all leaks repaired.

Step 8 – Clean Up the Job Site

Remove all scrap material around the outdoor unit and the furnace. Scrap material and debris such as chips of wood, pieces of tin, lengths of unused wire, dust from cleaning the filters, and bits of solder should be collected and placed in a proper trash receptacle. A neat installation increases the professional appearance of a job. Collect all tools and inventory them to determine that none are missing.

Step 9 – Instruct the Owner in Operation and Maintenance

Spend as much time as necessary to be sure that the owner feels confident in operating the system and understands routine owner maintenance. Most manufacturers provide an owner's operating and maintenance manual. Review this manual with the owner. In addition to demonstrating thermostat operation, the following are some of the more important points to discuss:

1. Keep the doors and windows closed.

2. Close the drapes to keep out direct sunlight.

3. Do not allow drapes to interfere with the free flow of air from the registers.

4. Clean grilles and coils in the outside unit as specified in the manufacturer's instructions.

5. Clean or change the furnace filter as required.
6. If the system does not cool properly, check and close the disconnect switch as required. Check and replace or reset the switch, and check the thermostat setting before requesting a service call.

All of the above actions are easily performed by the owner and can considerably reduce costly service calls.

GENERAL SERVICE PROBLEMS

If the system does not cool properly, follow the manufacturer's instructions to determine the condition and amount of refrigerant charge. This is usually done by analyzing refrigerant in the sight glass, by the pressure-temperature method, or by using a charging device and analyzing various temperatures. Add charge if necessary.

Refer to Units 15 and 16 of the text *Principles of Air Conditioning* for methods of verifying the accuracy of the duct size and unit size chosen for the cooling load. Units 11 and 12, regarding heating- and cooling-load estimating, may also be helpful.

REVIEW

Select the best answer.

1. Manufacturer's instructions are usually found
   a. in the service truck.
   b. taped or tied to the top of the unit.
   c. on the basement steps at the job site.
   d. underneath the slab.
2. After locating the manufacturer's installation instructions
   a. return them to the manufacturer.
   b. give them to the owner.
   c. study them carefully.
   d. put them in a safe place.
3. Unit-nameplate electrical phase and voltage should agree with
   a. the neighbors.
   b. the available power source.
   c. sound level ordinances.
   d. the last installation.
4. Most units operate best when installed on the side of the house that has
   a. the most grass.
   b. the least grass.
   c. a fence.
   d. the most shade at the hottest time of the day.
5. If the cooling coil and furnace are different brands
   a. they should not be installed in the same system.
   b. a flange arrangement may be required as a base to hold the coil.
   c. they should be painted the same color.
   d. they should be insulated.

6. When precharged tubing is used, it should be
   a. polished.
   b. cut to length.
   c. carefully curled in loops to take up the excess.
   d. shaped in S configurations to take up the excess.
7. All wiring should comply with
   a. the owner's wishes.
   b. the job requirements.
   c. the last installation.
   d. all local and national electrical codes.
8. When wiring a system, be sure to install ______________ on the wall near the outdoor unit.
   a. a rubber insulating pad
   b. a hook of adequate size
   c. a red flag
   d. a properly sized fused disconnect switch
9. If the system is not properly balanced after the first attempt
   a. open all doors for 5 minutes.
   b. install a larger system.
   c. readjust the registers until the proper airflow balance is reached.
   d. turn off the furnace fan.
10. Which of the following does not apply to the owner's routine operation and maintenance?
    a. Keep doors closed.
    b. Keep out direct sunlight.
    c. Clean grilles and filters.
    d. Charge refrigerant into the system regularly.

10
Installing Two-piece Residential Heat-pump Systems

OBJECTIVE

After studying this unit, the student will be able to

- install a central residential heat-pump air-conditioning system.

INTRODUCTION

A heat-pump system is very similar to the split system described in Unit 9. It operates in the same manner on the cooling cycle, and then reverses by thermostatic control for the heating cycle.

In this application, the outdoor heat-pump unit is connected to an indoor fan-coil unit, figure 10-1. The major difference between the outdoor unit of a heat pump and a cooling-only unit is the addition of a four-way valve that reverses the cooling cycle to a heating cycle.

Although installation, start-up, and service details may differ for each brand of heat pump, the general procedures presented here apply to most brands. Specific details are usually covered in the manufacturer's instructions available with each unit.

PROCEDURE

Step 1 – Getting Ready to Install the Outdoor Unit

1. Move the unit to the installation site and slide it out of the box, crate, or carton in which it is packaged.

2. Examine the unit for shipping damage such as bent panels, scratches, loose or bent fan blades or fan housing, crushed or bruised coil grilles, and cracked casings. File a claim if necessary.

3. Find the manufacturer's instructions taped or tied to the unit, in an envelope, in the shipping carton, or inserted in a convenient slot in the unit. Instructions are usually found taped or tied to the top of the unit.

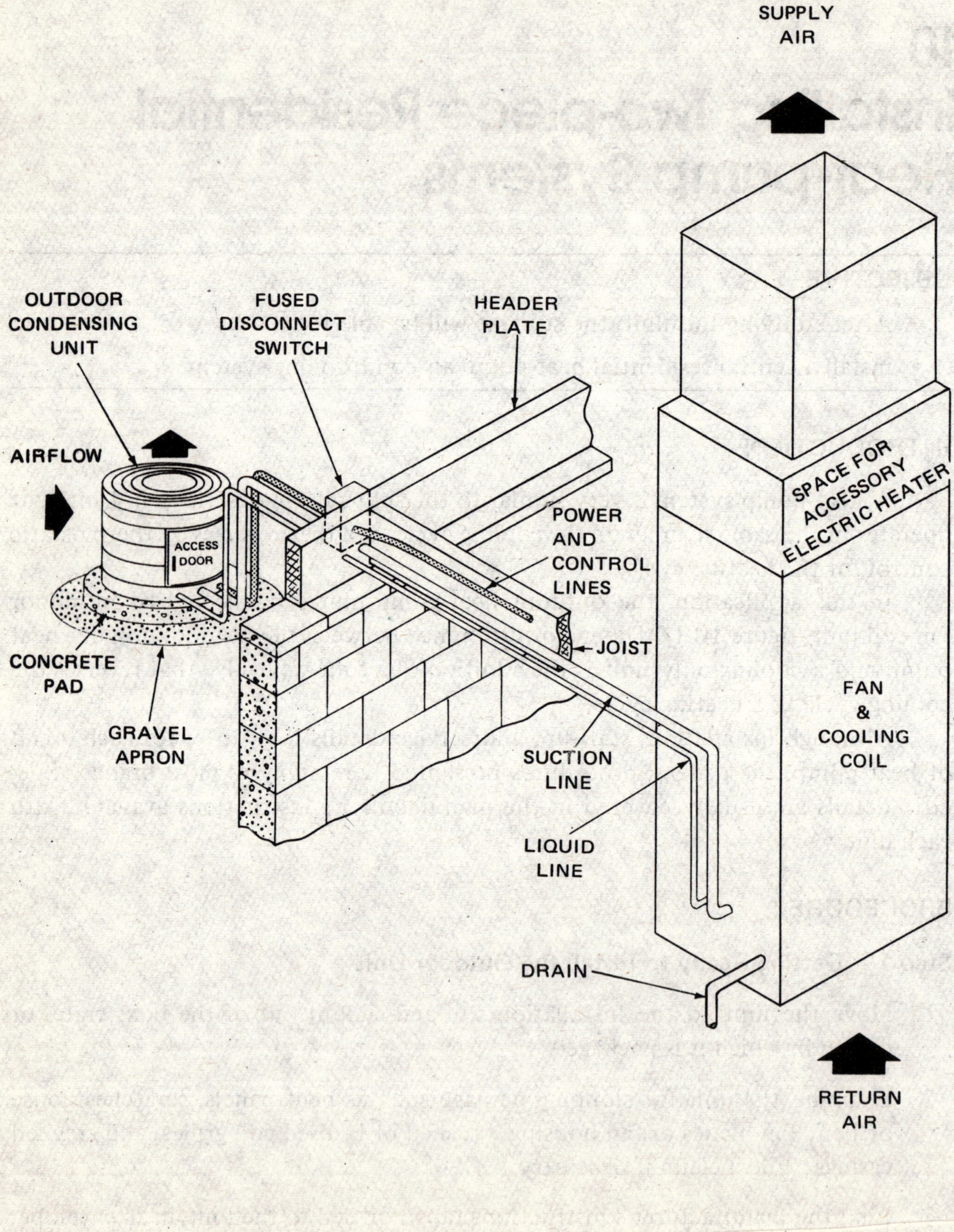

Fig. 10-1 Typical two-piece heat pump

4. Some manufacturers include a list of the items packed in the shipping carton in the front section of the instructions. If there is a list, carefully compare it with the items actually shipped, and report any missing parts. File a claim if necessary.

5. Study the manufacturer's instructions carefully and be sure you understand them.

6. Compare the Btu rating on the unit nameplate with the required cooling load and expected heating load for the space to be conditioned. The Btu capacity of the unit should be large enough to handle the cooling load and the partial-season heating load. In many areas, the heat pump can be sized large enough to handle the entire-season heating load.

7. Make certain that the nameplate electrical voltage and phase agree with the power source available.

8. Compare the space available for installation with the size of the unit and the space required for servicing.

9. Determine the location of the unit. Most units operate best during the cooling cycle when installed on the side of the house that has the most shade at the hottest time of day. The location of the outdoor unit must also take into account the unit's sound level. If the shadiest spot is on the property line and near neighbors' living areas, a compromise location should be selected, or an effective sound barrier such as a wall should be erected to comply with local noise ordinances.

Step 2 — Mount the Outdoor Base Pad and Condensing Unit

1. Install a concrete mounting pad on a gravel base located where rain or ice from the roof will not damage the unit. Allow sufficient space (consult the manufacturer's instructions) for future servicing and repair of the unit.
   a. Use a round pad for a round condensing unit, and a square pad for a square unit.
   b. The construction of the pad should be in proportion to the weight of the unit. Be sure the weight of the pad agrees approximately with the manufacturer's recommendations.
   c. Install tie-down bolts as indicated by the manufacturer. Two bolts are usually sufficient.
   d. Refer to the following (item 2) for spacing considerations.

   Note: If the unit is installed in a heavy snow area, follow the manufacturer's instructions for mounting the base pad on legs above the ground.

2. Place the condensing unit on the base pad. Be sure to allow enough space for making refrigerant piping runs and connections, and for running the power and control wiring to the terminals inside the unit. Enough clearance

for adequate airflow must also be allowed. If the service door or panel is positioned near a wall, allow sufficient space for easy access. Follow the manufacturer's instructions in all of these spacing considerations.

3. With the unit in place on the base pad, determine a suitable point for drilling holes in the header plate or foundation wall for the refrigerant piping and the control and power wiring, figure 10-2.

4. Slide the unit away from the wall (remove it from the base pad if necessary) and drill holes as required. The holes should be just large enough to accommodate the piping and wiring so that minimal caulking will be required later.

5. Reposition the unit and anchor it to the mounting bolts. If space is limited, it may be practical to install the wiring and the fused disconnect switches before anchoring the unit on the base pad.

Some manufacturers produce three-piece heat pumps consisting of: (1) an indoor fan and coil unit; (2) an outdoor coil without compressor or four-way valve (light enough to be mounted or hung on an outside wall); and (3) an indoor compressor and four-way valve unit that can be installed in the garage or other area where servicing may be more convenient. Wiring and piping circuits are essentially the same as for a two-piece heat pump. Follow the manufacturer's instructions for proper circuiting.

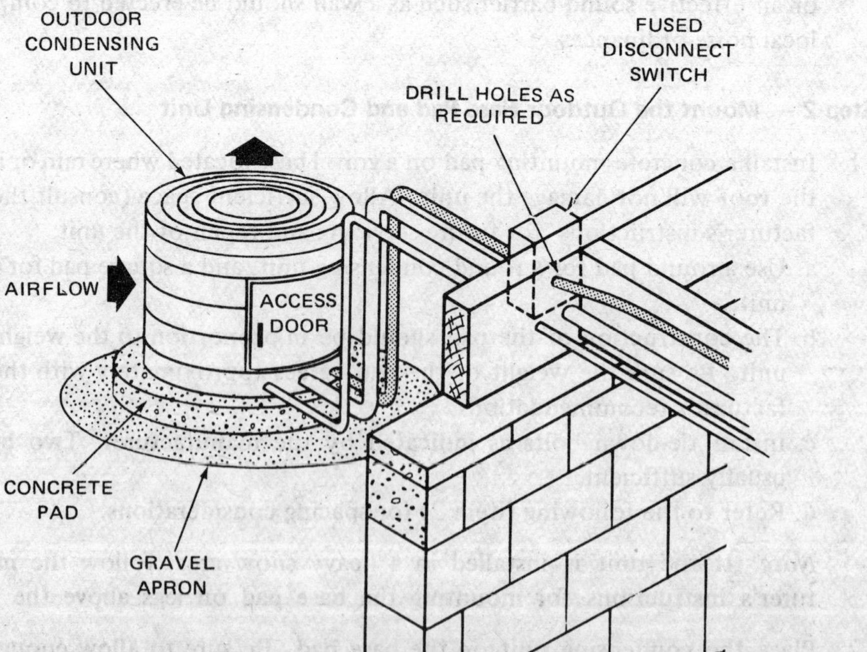

Fig. 10-2 Holes required for piping and wiring

Step 3 – Mount the Indoor Unit Coil and Fan

1. Move the coil and fan unit to the installation area.
2. Remove it from the box, crate, or carton and examine it for damage such as scratches, crushed fins, dented or bent panels, and loose mountings. File a claim if necessary.
3. Locate the manufacturer's instructions for the coil and fan unit and study them thoroughly.
4. Move the unit into its final position.
5. Install a properly sized disconnect switch for the fan motor.

Step 4 – Install the Refrigerant Piping

Quick-connect, speed-connect, compatible, flared, or sweat fittings can be used for making refrigerant piping connections. Some manufacturers supply piping or tubing packages of varying lengths precharged with refrigerant. These have quick, speed, or compatible fittings. Piping used with flared or sweat fittings is not precharged. If flared or sweat fittings are used, additional refrigerant should be added to the system according to the manufacturer's instructions.

Perform the following procedures when installing refrigerant piping:

1. Read the manufacturer's instructions carefully and determine the proper size and length of tubing for the equipment being installed. For this application, select a precharged tubing package as close to the required length as possible.
2. Do not cut precharged tubing. Carefully curl it into loops to take up any excess tubing. Even when tubing is not precharged, carefully examine the ends of the tubes before cutting them to the proper length. They may have been specially prepared to fit the connection, and will no longer do so if the end is removed.
3. Carefully unroll the tubing package, and run it from the condensing-unit connection through the holes in the header plate to the cooling-coil connection on the indoor unit. Keep in mind that with continuous tubing, bends should be made only by following acceptable tube-bending practices. Do not crimp or kink the tube by making a bend that is too sharp.
4. Install isolation hangers to prevent vibration noises from being transmitted to joists or beams.
5. Make quick-connect and compatible fitting connections at the condensing unit and cooling coil according to the manufacturer's instructions.
6. Install the drain line to an open drain.

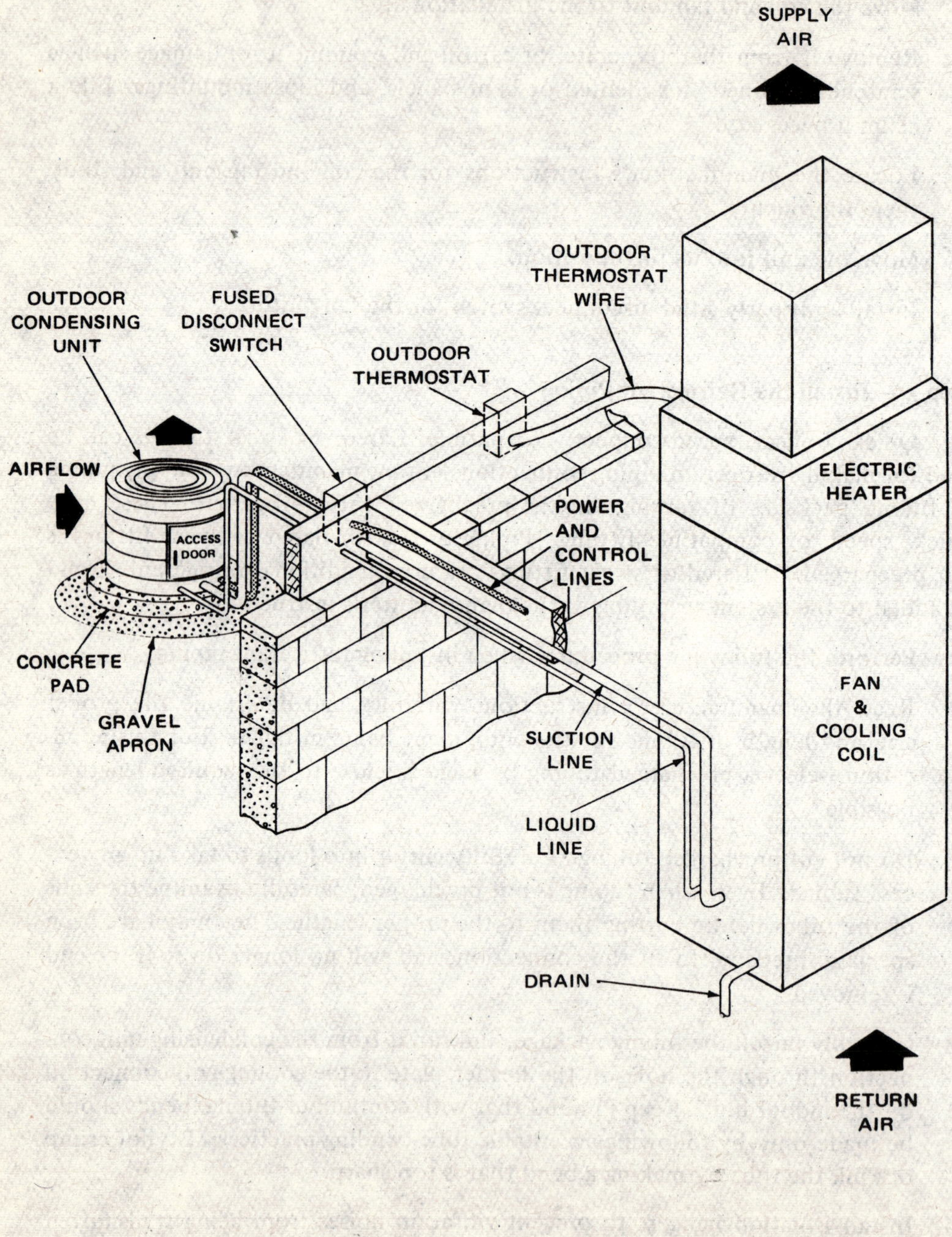

Fig. 10-3 Outdoor thermostat for accessory electric heater

Step 5 — Install the Wiring

All wiring runs, connections, and wire-size selections should be performed or closely supervised by an experienced, properly qualified electrician.

In this unit, the wiring instructions are based on the assumption that the compressor, fan, and motors are adequate for handling both heating and cooling requirements. A heating-cooling thermostat must be used.

If an accessory electric heater is used, follow the manufacturer's instructions to install an outdoor thermostat for the heater, figure 10-3.

Line Voltage Wiring to the Outdoor Unit. Perform the following operations when installing line voltage wiring to the compressor in the outdoor unit.

1. Study the manufacturer's instructions thoroughly and be certain that all wiring complies with local and national electrical codes and ordinances.
2. Install a properly sized fused disconnect switch on the wall near the outdoor unit.
3. Run line voltage wiring with the appropriate volt-ampere capacity (as specified in the manufacturer's instructions) from the power source to the outdoor disconnect switch.
4. Connect the wiring at the power source and at the disconnect switch, using proper safety precautions. Be sure the ground wire is connected at both ends and that the disconnect switches are in the OFF position at both the power source and the outdoor disconnect.
5. Run the power wiring, including ground wire, from the outdoor disconnect switch to the outdoor unit and connect it at the proper terminals, as indicated in the manufacturer's instructions.
6. Run the line voltage wiring to the indoor fan motor.

Low-voltage Control Wiring. To install low-voltage control wiring from the thermostat to the indoor and outdoor units, use the following general procedures:

1. Study the manufacturer's control-wiring instructions thoroughly and be sure you understand them.
2. Following the manufacturer's wiring instructions, run the low-voltage control wiring from the transformer to the indoor thermostat, and then to the proper indoor-unit contacts and the proper outdoor-unit contacts.
3. Install the heating-cooling thermostat.
4. Connect all control wiring to the terminals as indicated in the manufacturer's instructions.

Step 6 — Check Out All Piping and Wiring

Check all piping and wiring circuits and connections to make certain they agree with the manufacturer's instructions and with local and national electrical codes and ordinances.

Step 7 — Start Up the System

1. Open the suction and discharge valves according to the instructions to allow the free flow of refrigerant.
2. Close all disconnect switches to allow the free flow of electrical energy through the system components.
3. Check the registers in each room and adjust them according to the expected occupancy and coolness (or heat) required in each. In other words, open registers all the way in areas that are frequently and heavily occupied; open them halfway in areas that are used less often.
4. Set the thermostat to the desired temperature and allow the system to operate and settle out. This may take from several minutes to several hours, depending on the inside comfort load and the outside temperature.
5. If some areas do not reach the expected comfort temperature, readjust the registers. For example, move the wide-open registers to three-quarters open. This should cause more cool (or hot) air to flow into the areas that need it. Continue to adjust the registers in this manner until proper comfort levels are reached. If problems continue, refer to Unit 24, "Balancing the Air-conditioning System" of the text *Principles of Air Conditioning* for an expanded method of balancing the system.
6. Test the refrigerant system for leaks. This is usually done with an electronic tester, a halide tester, or soap bubbles. All joints should be examined and any leaks repaired.

Step 8 — Clean Up the Job Site

Remove all scrap material around the outdoor unit and the cooling coil. Scrap material and debris such as chips of wood, pieces of tin, lengths of unused wire, dust from cleaning the filters, and bits of solder should all be collected and placed in a proper trash receptacle. A neat installation increases the professional appearance of a job. Collect all tools and inventory them to determine that none are missing.

Step 9 — Instruct the Owner in Operation and Maintenance

Spend as much time as necessary to be sure that the owner feels confident in operating the system and understands routine owner maintenance. Most manufacturers provide an owner's operating and maintenance manual. Review this manual with the owner. In addition to demonstrating thermostat operation to the owner, discuss the following important procedures:

1. Keep the doors and windows closed (heating and cooling cycles).
2. Close the drapes to keep out direct sunlight (cooling cycle).
3. Do not allow drapes to interfere with the free flow of air from the registers (heating and cooling).
4. Clean grilles and coils on the outside unit as indicated in the manufacturer's instructions.
5. Clean or change filters as required (heating and cooling).
6. If the system does not heat or cool properly, check and close the disconnect switches as required. Check and replace the fuse or reset the fused switch, and check the thermostat setting before requesting service.

All of the above actions are easily performed by the owner and can considerably reduce costly service calls.

GENERAL SERVICE PROBLEMS

If the system does not operate properly, follow the manufacturer's instructions to determine the condition and amount of refrigerant charge. This is usually done by analyzing refrigerant in the sight glass, by the pressure-temperature method, or by using a charging device and analyzing various temperatures. Add charge if necessary.

Refer to Units 15 and 16 of the text *Principles of Air Conditioning* for methods of verifying the accuracy of the duct sizes and unit size chosen for the cooling and heating loads. If the ducts are too small, airflow to the space may not be enough to maintain adequate comfort. If the unit is too small, adequate cooling or heating may not be possible. Units 11 and 12 of *Principles of Air Conditioning*, regarding heating- and cooling-load estimating, may also be helpful in verifying the comfort load.

REVIEW

Select the best answer.

1. Manufacturer's instructions are usually found
   a. in the service truck.
   b. taped or tied to the top of the unit.
   c. on the basement steps at the job site.
   d. underneath the slab.

2. After locating the manufacturer's installation instructions
   a. return them to the manufacturer.
   b. give them to the owner.
   c. study them carefully.
   d. put them in a safe place.

3. Unit nameplate electrical phase and voltage should agree with
   a. the neighbors.
   b. the available power source.
   c. sound level ordinances.
   d. the last installation.

4. Most units operate best on the cooling cycle when installed on the side of the house that has
   a. the most grass.
   b. the least grass.
   c. a fence.
   d. the most shade at the hottest time of the day.

5. A heat-pump system may require
   a. an indoor thermostat only.
   b. an outdoor thermostat only.
   c. an indoor and an outdoor thermostat.
   d. no thermostat because it works off ambient air.

6. When precharged tubing is used it should be
   a. polished.
   b. cut to length.
   c. carefully curled in loops to take up the excess.
   d. shaped in S configurations to take up the excess.

7. All wiring should comply with
   a. the owner's wishes.
   b. the job requirements.
   c. the last installation.
   d. all local and national electrical codes.

8. When wiring a system, be sure to install ________________ on the wall near the outdoor unit.
   a. a rubber insulating pad
   b. a hook of adequate size
   c. a red flag
   d. a properly sized fused disconnect switch

9. If the system is not properly balanced after the first attempt
   a. open all doors for 5 minutes.
   b. install a larger system.
   c. readjust the registers until the proper airflow balance is reached.
   d. turn off the fan.

10. Which of the following does not apply to the owner's routine operation and maintenance?
    a. Keep doors closed.
    b. Keep out direct sunlight.
    c. Clean grilles and filters.
    d. Charge refrigerant into the system regularly.

11
Installing Gas-fired Residential Furnaces

OBJECTIVES

After studying this unit, the student will be able to

- install a residential gas-fired furnace.
- discuss the need to construct ducts which provide adequate air for combustion.

INTRODUCTION

A central residential furnace is designed to be connected to a duct system that is capable of delivering air to all rooms in the house simultaneously. An upflow furnace installed in the basement, figure 11-1, connects to a duct system in the basement ceiling. Installed on the main floor in a house without a basement, an upflow furnace connects to a duct system above the ceiling. A downflow furnace on the main floor connects to a duct system in the space under the floor, figure 11-2.

The furnace described here is a gas-fired, upflow model installed in the basement. It burns natural gas to produce heat. Air flows through it from the bottom (return air) to the top (supply air) of the furnace. This furnace does not have a wasteful standing gas pilot light, it uses an electric ignitor that provides spark to light the gas pilot when the burner needs to be ignited. The installation procedure for a downflow furnace is similar to that of an upflow furnace, except that air flows through the furnace from the top (return air) to the bottom (supply air).

Installation, start-up, and service details may differ with each manufacturer, but the general procedures outlined here apply to most brands. Specific details are usually covered in the manufacturer's instructions available with each unit.

PROCEDURE

Step 1 – Getting Ready to Install the Furnace

1. Move the unit to the installation site and slide it out of the box, crate, or carton in which it is packaged.
2. Examine the unit for shipping damage such as bent panels, scratches, loose or bent fans or fan housing, and cracked casings, and file a claim if necessary.

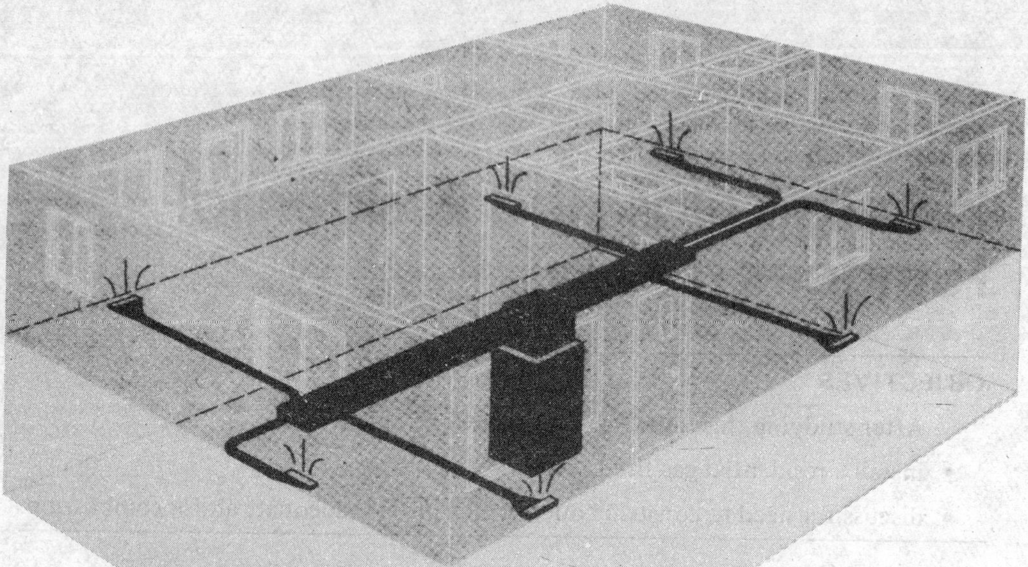

Fig. 11-1 Upflow furnace

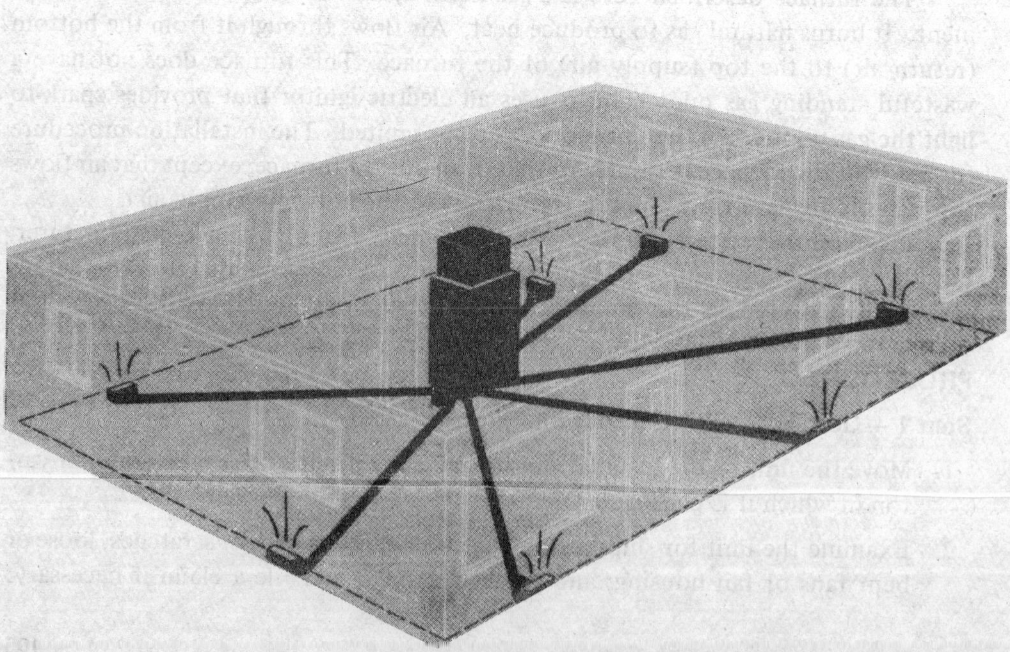

Fig. 11-2 Downflow furnace

3. Find the manufacturer's instructions taped or tied to the unit, in an envelope, in the shipping carton, or inserted in a convenient slot in the unit. Most instructions are found taped or tied to the top of the unit.
4. Some manufacturers include a list of the items packed in the shipping carton in the front section of the instructions. If there is a list, carefully compare it with the items actually shipped and report any missing parts. File a claim if necessary.
5. Study the manufacturer's instructions carefully and be sure you understand them.
6. Compare the Btu rating on the unit nameplate with the required heating load for the space to be conditioned. The Btu capacity of the unit should be large enough to handle the heating load.
7. Make certain that the nameplate electrical voltage and phase agree with the power source available.
8. Compare the space available for installation with the size of the furnace and the space required for servicing as indicated by the manufacturer.
9. Make certain that an appropriate gas supply is available.

Step 2 – Make Duct Connections to the Furnace

1. Connect the furnace plenum to the supply air plenums in the duct system, figure 11-3. Use a standard duct-installation procedure.
2. Connect the return air duct plenum to the return air connection on the furnace, figure 11-4. If a filter is to be installed, be sure to allow enough space at the return air plenum connection to the furnace. (Filter installation is discussed in detail in Unit 12.)

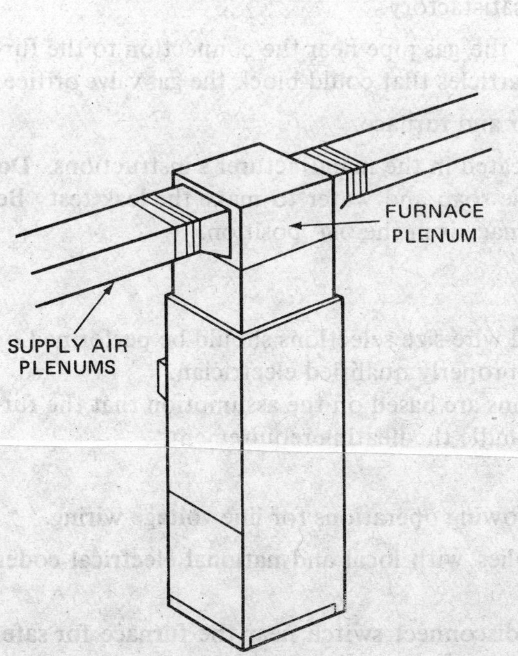

Fig. 11-3 Connect furnace plenum to supply air ducts

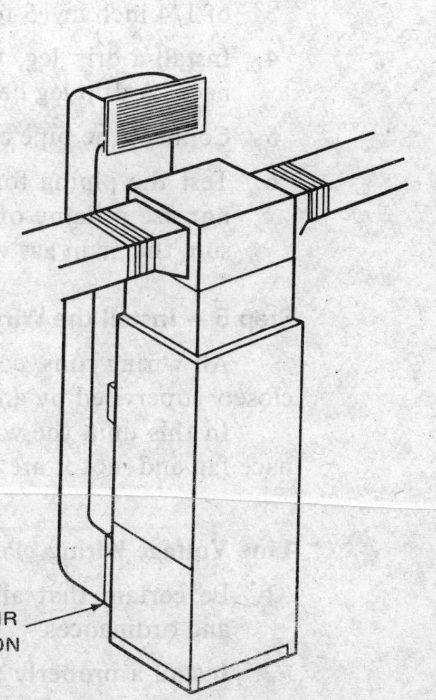

Fig. 11-4 Connect return air duct to furnac

Step 3 – Install the Furnace Flue

When installing a furnace flue pipe, follow standard procedures that comply with all local and national safety codes. The following items should be used as general guidelines.

1. Locate the furnace as close to the chimney as possible so the flue pipe is as short as possible. A short flue reduces the number of joints required and therefore minimizes potential leakage points, figure 11-5.
2. The full length of the flue should be as large as the furnace outlet and should run in an upward slope from the furnace to the chimney.
3. Flue pipe material must comply with local codes.
4. Always connect the furnace flue to a furnace-only chimney. **Never connect it to a fireplace chimney.**
5. The furnace-only chimney must be properly installed and have sufficient roof clearance for proper draft. Follow local code requirements and the manufacturer's instructions.

Step 4 – Install the Gas Piping

1. Use standard gas piping practices, following local and national codes and ordinances.
2. Select the pipe size as specified in the manufacturer's instructions and National Fire Protection Association (NFPA) requirements. The pipe should be large enough to handle the maximum cubic feet of gas required by the furnace.
3. Run the pipe sloping downward away from the meter to the furnace. A slope of 1/4 inch in 15 feet should be satisfactory.
4. Install a drip leg, figure 11-6, in the gas pipe near the connection to the furnace. A drip leg helps trap dirt particles that could block the gas valve orifice.
5. Connect the pipe at the gas meter and furnace.
6. Test the piping for leaks as indicated in the manufacturer's instructions. **Do not use a flame of any kind.** Use soap and water to make the leak test. Be sure the main gas valve at the furnace is in the OFF position.

Step 5 – Install the Wiring

All wiring runs, connections, and wire-size selections should be performed or closely supervised by an experienced, properly qualified electrician.

In this unit, the wiring instructions are based on the assumption that the furnace fan and motor are adequate to handle the heating requirements.

Line Voltage Wiring. Perform the following operations for line voltage wiring.

1. Be certain that all wiring complies with local and national electrical codes and ordinances.
2. Install a properly sized furnace disconnect switch near the furnace for safe, easy shutoff when servicing.

Fig. 11-5 Short flue pipe

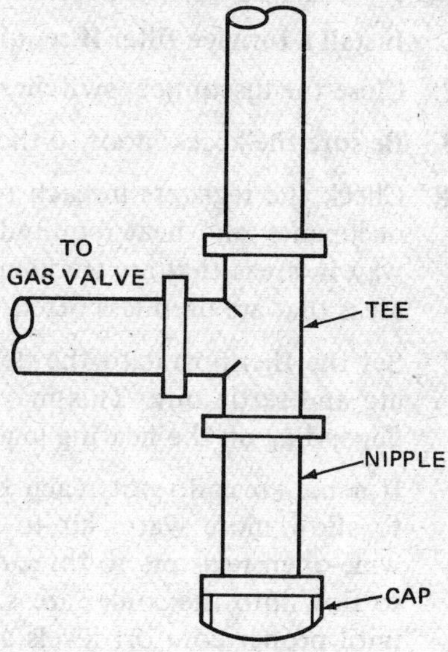

Fig. 11-6 Drip leg

3. Run the appropriate line voltage wiring from the power source to the furnace disconnect switch. Follow the manufacturer's instructions.
4. Run wiring from the disconnect switch to the furnace.
5. Be sure the ground wire is connected at both ends and that disconnect switches are in the OFF position at both the power source and the furnace. Connect the wiring at the power source and at the disconnect switch, using proper safety precautions.

Low-voltage Control Wiring. To install low-voltage control wiring from the thermostat to the furnace, use the following general procedures:

1. Study the manufacturer's control-wiring instructions thoroughly and be sure you understand them.
2. Shut off any power to the furnace.
3. Install a subbase for the heating thermostat.
4. Following the manufacturer's wiring instructions closely, run low-voltage control wiring from the furnace or control-box transformer to the proper thermostat subbase controls.
5. Install the heating thermostat.
6. Connect all control wiring to the proper terminals as shown in the manufacturer's instructions.

Step 6 — Check Out All Piping and Wiring

Check all piping and wiring circuits and connections to make certain they agree with the manufacturer's instructions and with local and national gas and electrical codes and ordinances, especially NFPA codes.

Step 7 – Start Up the Furnace

1. Install a furnace filter if required.
2. Close the disconnect switches.
3. Be sure the access door to the fan area is closed.
4. Check the registers in each room and adjust them according to the expected occupancy and heat required in each. In other words, open registers all the way in areas that are frequently and heavily occupied; open them halfway in areas that are used less often.
5. Set the thermostat to the desired temperature and allow the system to operate and settle out. This may require from several minutes to several hours, depending on the heating load and the outside temperature.
6. If some areas do not reach the expected temperature, readjust the registers to allow more warm air to enter the colder areas. For example, move the wide-open registers to three-quarters open. This should cause more warm air to flow into the colder areas. Continue to adjust the registers in this manner until proper comfort levels are reached. If problems continue, refer to Unit 24, "Balancing the Air-conditioning System" of the text *Principles of Air Conditioning* for an expanded method of balancing the air flow.

Step 8 – Clean Up the Job Site

Remove all scrap material around the thermostat, and the furnace. Scrap material and debris such as chips of wood, pieces of tin, lengths of unused wire, dust, and bits of wood should be collected and either saved or placed in a proper trash receptacle. A neat installation increases the professional appearance of a job. Collect all tools and inventory them to determine that none are missing.

Step 9 – Instruct the Owner in Operation and Maintenance

Spend as much time as necessary to be sure that the owner feels confident in operating the system and understands routine owner maintenance. Most manufacturers provide an owner's operating and maintenance manual. Review this manual with the owner. In addition to demonstrating thermostat operation to the owner, discuss the following important procedures:

1. Keep the doors and windows closed.
2. Open the drapes to allow direct sunlight to enter the room.
3. Do not allow drapes to interfere with the free flow of air from the registers.
4. Clean or change the filter in the furnace as required.
5. If the system does not heat properly, check and close the disconnect switch as required. Check and replace or reset the fused disconnect switch. Check for proper thermostat setting before requesting service.

All of the above actions are easily performed by the owner and can considerably reduce costly service calls.

GENERAL SERVICE PROBLEMS

If the system does not heat properly, follow the manufacturer's troubleshooting sequence. Verify the duct system for proper duct size; refer to Unit 15 of the text *Principles of Air Conditioning*. Verify the total heating load to determine proper furnace size; refer to Units 11 and 12, *Principles of Air Conditioning*.

If the furnace fan motor has a two-speed capacity, modify the terminal connections according to the manufacturer's instructions to increase airflow by increasing fan speed. If problems persist, refer to Units 15 and 16 of *Principles of Air Conditioning* for methods of verifying accuracy of the heating-load estimate and the size of the ducts and furnace chosen for handling the heating load. If the ducts or the unit are too small, supplying adequate heat to the space may not be possible.

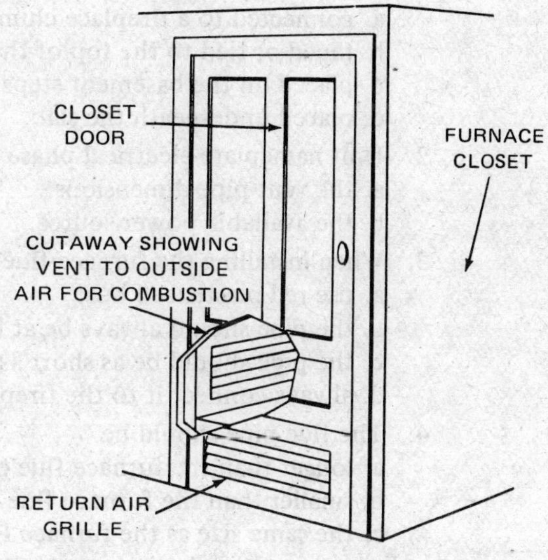

Fig. 11-7 Adequate supply of combustion air

COMBUSTION AIR

Because of tighter building construction and the increased loss of air through various exhaust fans, fireplaces, clothes dryers, etc., many newer homes require special consideration to assure adequate air for furnace combustion.

Former building practices resulted in a looser construction that permitted sufficient air for blue-flame combustion to infiltrate the building. Tighter construction and more weatherstripping and caulking have all but eliminated that source of air. To offset this condition, direct ducting to the outdoors for adequate furnace combustion-chamber air is recommended, figure 11-7.

To determine the proper location, size, and construction of a duct for combustion air, study local safety and fire-prevention codes and ordinances. Some manufacturers have described in great detail the safe and effective application of their products. Be sure to follow the manufacturer's instructions for proper installation of these products and related combustion air ducts.

Venting the Furnace

Proper vent pipes to carry the products of combustion to the outdoors are also essential. Again, local codes and ordinances should be followed to insure safe and properly designed vent pipes. In most cases, the vent pipe should extend a specified distance beyond the roof line.

Follow the vent manufacturer's installation instructions.

REVIEW

Select the best answer.

1. A furnace flue pipe should never be
   a. connected to a fireplace chimney flue.
   b. taped or tied to the top of the unit.
   c. placed on the basement steps at the job site.
   d. placed underneath the slab.
2. Unit nameplate electrical phase and voltage should agree with
   a. the vent-pipe dimensions.
   b. the available power source.
   c. sound-level ordinances.
   d. the last installation.
3. When installing the furnace flue
   a. use red plasticized hose.
   b. the pipe should always be at least 10 feet long.
   c. the pipe should be as short as possible to reduce the number of joints.
   d. always connect it to the fireplace chimney flue.
4. The flue pipe should be
   a. longer than the furnace flue outlet.
   b. smaller than the furnace flue outlet.
   c. the same size as the furnace flue outlet.
   d. wrapped in plastic to make it airtight.
5. The flue pipe from the furnace to the chimney should slope
   a. downward.
   b. upward.
   c. to the left.
   d. to the right.
6. All wiring should comply with
   a. the owner's wishes.
   b. the job requirements.
   c. the last installation.
   d. all local and national electrical codes.
7. When wiring the system, be sure to install ________________ on the wall near the furnace.
   a. a rubber insulating pad
   b. a hook of adequate size
   c. a red flag
   d. a properly sized disconnect switch
8. If the system is not properly balanced after the first attempt
   a. open all the doors for 5 minutes.
   b. install a larger system.
   c. readjust the registers until proper airflow balance is reached.
   d. turn off the furnace fan.
9. Which of the following does not apply to the owner's routine operation and maintenance?
   a. Keep doors closed.
   b. Allow direct sunlight to come in the windows.
   c. Clean the furnace filters.
   d. Change the furnace fan motor.
10. Sufficient air for blue-flame combustion can be achieved by
    a. constructing a vent to the outside.
    b. careful weatherstripping and caulking.
    c. running the clothes dryer frequently.
    d. opening a window at night.

12 Installing Electronic Air Cleaners

OBJECTIVE

After studying this unit, the student will be able to install an electronic air cleaner with an upflow gas furnace.

INTRODUCTION

Electronic air cleaners may be installed with upflow, downflow, or horizontal-flow heating or cooling systems. In each application the general procedures are similar; only the location of the cleaner changes. For example, with an upflow furnace, the air cleaner is installed at the return air connection at either the side or the bottom of the furnace, figure 12-1. With a downflow furnace, the cleaner is located at the top of the furnace at the return air connection. With a horizontal-flow furnace, the cleaner is located in the return air plenum, using a duct transition fitting as required.

When installed with cooling units, the cleaner is usually located ahead of the cooling coil in the return air ductwork to minimize moisture at the cleaner.

PROCEDURE

Step 1 – Getting Ready to Install the Electronic Air Cleaner

1. Move the unit to the installation site and slide it out of the box, crate, or carton in which it is packaged.
2. Examine the air cleaner for shipping damage such as bent panels, scratches, loose or broken wires and connectors, and file a claim if necessary.
3. Find the manufacturer's instructions taped or tied to the unit, in an envelope, or in the shipping carton. Most instructions are found taped to the top of the unit.

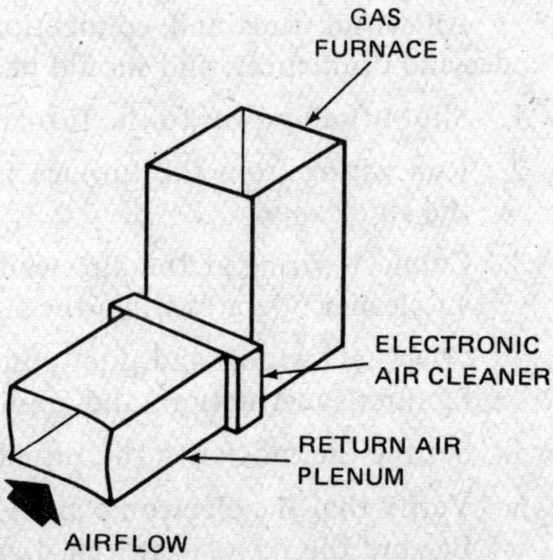

Fig. 12-1 Typical electronic air cleaner with upflow gas furnace

4. Some manufacturers include a list of items shipped at the beginning of the installer's book. If there is a list, carefully compare it with the items actually shipped. Report any missing parts, and file a claim if necessary.
5. Study the manufacturer's installation instructions carefully and be sure you understand them.
6. Compare the nameplate electrical ratings with the available power source. Be certain that they agree.
7. Compare the space available at the return air connection to the furnace with the size of the air cleaner.

Step 2 – Prepare the Furnace and Install the Air Cleaner

Most electronic air cleaners are shipped with a template with instructions for preparing the furnace.

1. Follow the template instructions for cutting an appropriate opening in the furnace for the cleaner. Prepare the opening according to the instructions.
2. Prepare the return air duct. If the duct is larger or smaller than the air cleaner connection, use a transition fitting for smoother airflow.
3. Position the air-cleaner cabinet and connect it to the furnace and return air duct. Be sure all connections are airtight and properly sealed.
4. Install the various cleaner cells in the cleaner cabinet as shown in the manufacturer's instructions. If there are airflow arrows on the cells, be certain they point in the same direction as the airflow through the unit. When a prefilter is used, it should be installed ahead of the air-cleaner cells.

Step 3 – Install Wiring to the Electronic Air Cleaner

All wiring runs and connections should follow local and national electrical codes and ordinances, and should be performed by a properly qualified electrician.

1. Shut off all power to the furnace fan motor.
2. Run wiring from the furnace fan motor to the power-supply connections on the air cleaner.
3. Connect wiring at the air cleaner and at the furnace fan-motor circuit so the air cleaner operates when the furnace fan motor operates.
4. Check all wiring and duct connections to the air cleaner against the manufacturer's instructions and applicable codes.
5. Set the thermostat so the furnace fan motor operates.
6. Verify that the electronic air cleaner operates when the fan motor is running. Be sure the ON-OFF switch, figure 12-2, is ON and then test the operation of the unit according to the manufacturer's instructions. A test button and an indicator light are usually provided to test the unit's operation, figure 12-3.
7. If the unit does not operate properly, follow the manufacturer's troubleshooting instructions to locate and correct the problem.

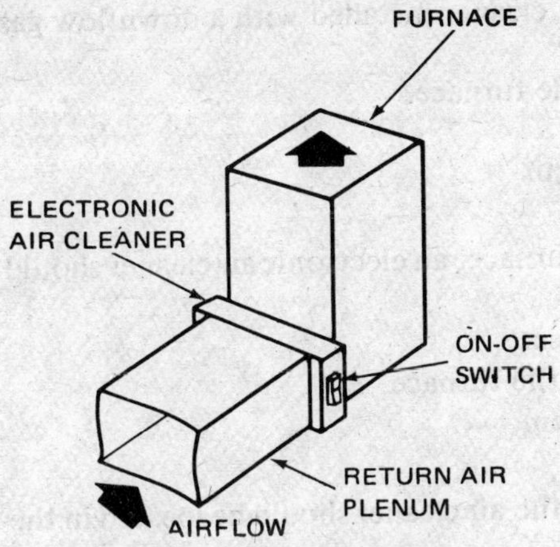

Fig. 12-2 Air cleaner ON-OFF switch

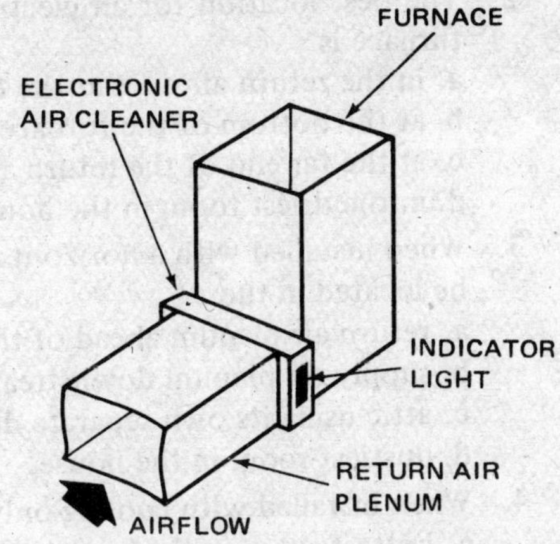

Fig. 12-3 Indicator light on air cleaner

Step 4 – Clean Up the Job Site

Remove all scrap material and debris from around the installation, and either save or deposit it in a proper trash receptacle. Collect and inventory all tools.

Step 5 – Instruct the Owner in Operation and Maintenance

Most manufacturers provide an owner's operating and maintenance manual. Review this manual with the owner until he or she feels confident in operating and maintaining the air cleaner. Some key points to discuss are:

1. The unit must be switched to ON so it can operate when the furnace fan is running.
2. The unit should operate properly when the indicator light (if available) is glowing.
3. Switch the unit to OFF and switch the disconnect switch to the unit circuit to the OFF position, before opening the cleaner or before removing the electronic cells or prefilter.
4. Switch the unit to OFF to stop the operation.
5. Make certain that the unit is turned on and that the thermostat is set at a temperature that causes the furnace fan to operate before requesting service.

REVIEW

Select the best answer.

1. The best location for an electronic air cleaner, when installed with an upflow gas furnace, is
   a. in the return air duct at the bottom or lower side of the furnace.
   b. at the top of the furnace.
   c. at the far end of the return air plenum.
   d. at the far end of the supply air plenum.

2. The best location for an electronic air cleaner installed with a downflow gas furnace is
   a. in the return air duct at the top of the furnace.
   b. at the bottom of the furnace.
   c. at the far end of the return air plenum.
   d. in the driest room in the house.
3. When installed with a horizontal-flow furnace, an electronic air cleaner should be located in the
   a. return air plenum ahead of the furnace.
   b. supply air plenum downstream from the furnace.
   c. attic using its own separate duct system.
   d. dustiest room in the house.
4. When installed with cooling-only units, the air cleaner should be located in the
   a. hottest room in the house.
   b. dustiest room in the house.
   c. return air plenum ahead of the cooling coil.
   d. supply air plenum where it can be kept constantly moist.
5. Most electronic cleaners are shipped with
   a. insulation.
   b. sheet metal tape.
   c. nuts and bolts.
   d. a template with instructions for preparing the opening in the furnace.
6. If the air cleaner is smaller than the return air duct plenum
   a. use tape at the connection.
   b. use a gradual duct transition for smooth airflow.
   c. use a box connection to minimize duct length.
   d. use C clamps to make the joint permanent.
7. All air-cleaner duct connections should be
   a. painted red.
   b. painted blue.
   c. waxed.
   d. airtight.
8. Any airflow arrows on the air cleaner cells should be pointing
   a. up.
   b. down.
   c. downstream.
   d. in the direction of the airflow in the return air duct.
9. An electronic air cleaner should operate
   a. only after the humidity level is accurate.
   b. when the furnace fan operates.
   c. only during daytime hours.
   d. when the air is hottest.
10. Usually, the owner can determine that the air cleaner is operating
    a. by observing that the air cleaner indicator light is glowing when the furnace fan is running.
    b. when a steady tone is heard.
    c. by feeling a sudden vibration.
    d. if furniture dusting is not required for 36 hours.

13
Installing Central Residential Humidifiers

OBJECTIVE

After studying this unit, the student will be able to

- install a humidifier in a central residential duct system.

INTRODUCTION

Basic, central-system humidifiers include duct, bypass, and fan-powered models. A duct humidifier, figure 13-1, is usually located in the supply air plenum; bypass and fan-powered types are attached to the furnace plenum. In each of these applications, the general installation procedures are similar. Each must be located either directly in the system airstream or have direct access to the system air. Each must have a controlled water supply, and each must be cleaned periodically. Power-drive and fan-powered models also require an electric power source.

PROCEDURE

Step 1 – Getting Ready to Install the Humidifier

1. Move the humidifier to the installation site and slide it out of the box, crate, or carton in which it was shipped.
2. Examine the humidifier for any shipping damage such as dents, scratches, and cracks, and file a claim if necessary.
3. Find the manufacturer's instructions taped or tied to the unit or in an envelope.
4. Look for a list of items shipped. This is usually located at the beginning of the installer's book. If there is a list, compare the items shipped with the items on the list. Report any missing parts and file a claim if necessary.
5. Study the manufacturer's installation instructions carefully and be sure you understand them.
6. Compare the electric power requirements of the humidifier with the power supply.
7. Compare the size of the humidifier with the size of the supply air plenum duct and surface.

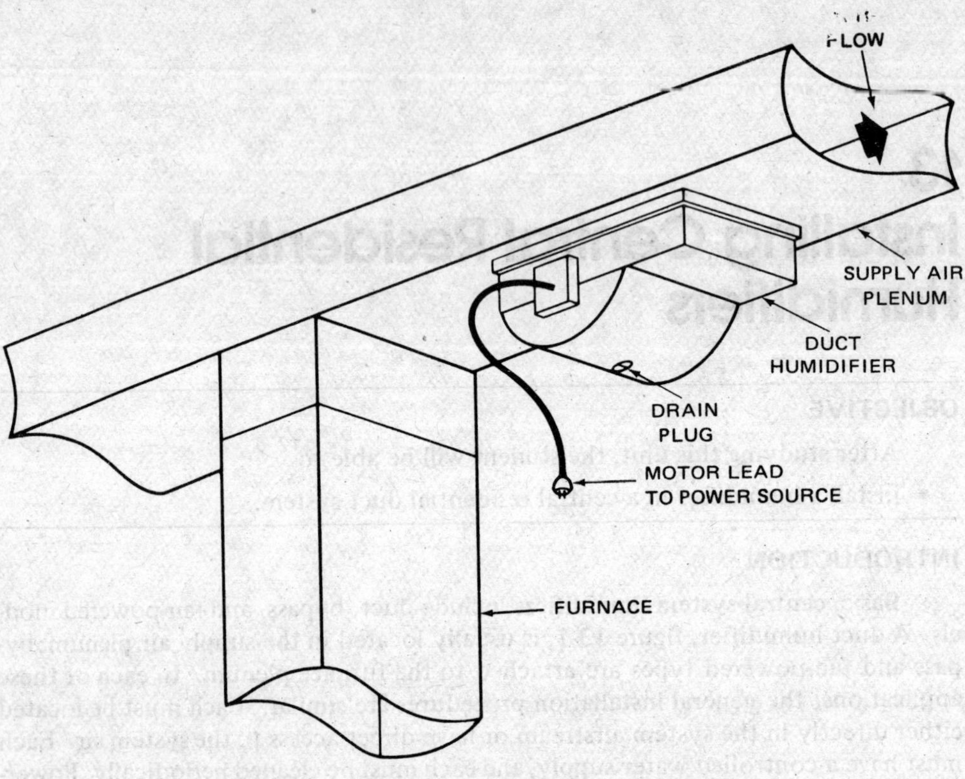

Fig. 13-1 Typical duct humidifier installation

Step 2 – Prepare the Supply Air Plenum Duct and Install the Humidifier

Most humidifiers are supplied with a template containing instructions for preparing an opening in the supply air duct.

1. Follow the template instructions for cutting an opening in the supply duct to receive the humidifier. Prepare the opening according to the manufacturer's installation instructions.
2. Position the humidifier and secure it to the plenum duct, as shown in the manufacturer's instructions.

Step 3 – Install the Water Line to the Humidifier

1. Shut off the main water supply and run a proper sized tube, according to the manufacturer's instructions, from the main cold water pipe to the humidifier. The tubing should include a saddle shutoff valve.
2. Connect the water tube to the humidifier and water pipe according to the manufacturer's instructions.
3. Close the saddle shutoff valve in the water tubing line to the humidifier. The valve should remain closed until all wiring and duct connections have been checked (as explained in Step 4, item 5).

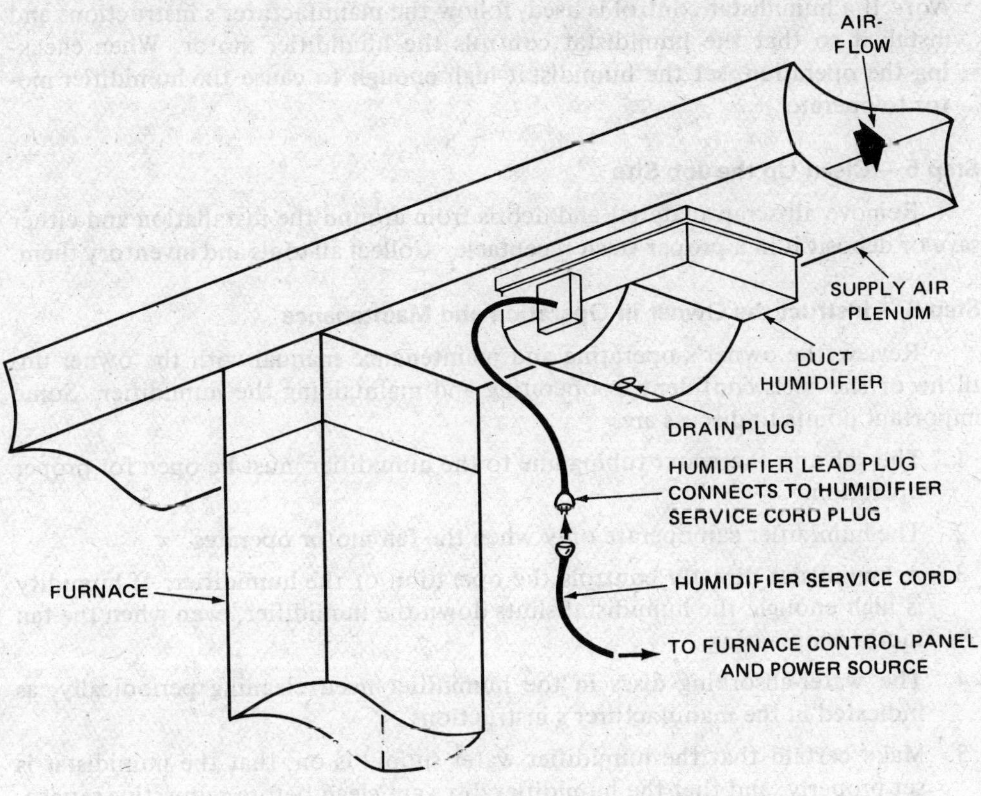

Fig. 13-2 Connect humidifier lead plug to service cord

4. Open the main water valve so that water is available at the humidifier shutoff valve.

Step 4 – Install the Wiring and Check Humidifier Operation

Follow all local and national electrical codes and ordinances.

1. Shut off the power supply to the furnace fan motor control panel.
2. Run wiring from the power source and furnace fan motor control panel circuit to the humidifier service cord leads.
3. Connect the humidifier service cord wiring leads so the humidifier operates when the furnace fan motor operates.
4. Connect the humidifier plug to the humidifier service cord plug, figure 13-2.
5. Check to insure that all wiring, duct connections, and water connections agree with the manufacturer's instructions and applicable codes.
6. Open the valve in the water tubing line to the humidifier.
7. Turn on the power to the furnace fan motor control panel circuit.
8. Set the thermostat so the furnace fan motor operates.
9. Verify that the humidifier motor operates when the fan motor runs.

Note: If a humidistat control is used, follow the manufacturer's instructions and install it so that the humidistat controls the humidifier motor. When checking the operation, set the humidistat high enough to cause the humidifier motor to operate.

Step 5 – Clean Up the Job Site

Remove all scrap material and debris from around the installation and either save or deposit it in a proper trash receptacle. Collect all tools and inventory them.

Step 6 – Instruct the Owner in Operation and Maintenance

Review the owner's operating and maintenance manual with the owner until he or she feels confident in operating and maintaining the humidifier. Some important points to discuss are:

1. The valve in the water tubing line to the humidifier must be open for proper operation.
2. The humidifier can operate only when the fan motor operates.
3. A humidistat directly controls the operation of the humidifier. If humidity is high enough, the humidistat shuts down the humidifier, even when the fan motor is operating.
4. The water-absorbing discs in the humidifier need cleaning periodically, as indicated in the manufacturer's instructions.
5. Make certain that the humidifier water supply is on, that the humidistat is set properly, and that the humidifier discs are clean before requesting service.

REVIEW

Select the best answer.

1. A duct humidifier is usually located in
   a. the supply air plenum.
   b. the return air plenum.
   c. the furnace combustion chamber.
   d. the attic.
2. Humidifiers must have
   a. a controlled water supply.
   b. a glass window.
   c. a block water reservoir.
   d. two inches of insulation.
3. Compare humidifier space requirements with the
   a. furnace size.
   b. basement size.
   c. supply air plenum size.
   d. flue pipe size.
4. Follow template instructions for
   a. cutting into the main water supply.
   b. positioning the supply duct.
   c. cutting an opening in the supply air duct.
   d. positioning the furnace wall.
5. Before connecting the humidifier water tubing to the main cold-water supply
   a. plug in the electricity.
   b. open the furnace gas valve.
   c. plug the drain.
   d. shut off the main water supply.

6. The saddle shutoff valve in the humidifier water tubing line should be closed until
   a. the water temperature reaches 68 F.
   b. the reservoir is full.
   c. the humidifier is plugged.
   d. all wiring, duct, and water connections have been completed and checked.
7. Before running wiring and making electrical connections
   a. clean the furnace.
   b. polish the humidifier.
   c. shut off all power to the furnace fan motor control board.
   d. buy 2 rolls of tape.
8. Open the saddle valve in the humidifier water tubing line
   a. after checking the wiring, duct, and humidifier water-line connections.
   b. early each morning.
   c. one half turn.
   d. one eighth turn.
9. The humidifier motor should be wired to operate
   a. three times per day.
   b. three times per hour.
   c. only when the furnace fan motor operates.
   d. early in the morning.
10. Before calling for humidifier service, the owner should
    a. make certain the humidifier water supply is on.
    b. close all drapes.
    c. tap the humidifier 3 times.
    d. wait for dry, sunny weather.

ANSWERS TO REVIEW

Unit 1

1. j
2. b
3. i
4. g
5. c
6. d
7. e
8. f
9. h
10. a

Unit 2

1. d
2. c
3. d
4. d
5. b
6. c
7. d
8. d
9. c
10. a

Unit 3

1. c
2. c
3. a
4. c
5. c
6. c
7. d
8. c
9. c
10. c

Unit 4

1. c
2. a
3. b
4. d
5. a
6. d
7. d
8. b
9. d
10. a

Unit 5

1. c
2. b
3. c
4. b
5. b
6. c
7. a
8. b
9. c
10. b

Unit 6

1. c
2. c
3. c
4. d
5. b
6. c
7. d
8. a
9. b
10. c

Unit 7

1. b
2. c
3. d
4. c
5. c
6. a
7. b
8. d
9. a
10. d

Unit 8

1. b
2. c
3. d
4. b
5. c
6. b
7. d
8. c
9. a
10. b

Unit 9

1. b
2. c
3. b
4. d
5. b
6. c
7. d
8. d
9. c
10. d

Unit 10

1. b
2. c
3. b
4. d
5. c
6. c
7. d
8. d
9. c
10. d

Unit 11

1. a
2. b
3. c
4. c
5. b
6. d
7. d
8. c
9. d
10. a

Unit 12

1. a
2. a
3. a
4. c
5. d
6. b
7. d
8. d
9. b
10. a

Unit 13

1. a
2. a
3. c
4. c
5. d
6. d
7. c
8. a
9. c
10. a